Capt. Syd Jones

Atocha Treasure Adventures:

Sweat of the Sun, Tears of the Moon

Atocha Treasure Adventures: Sweat of the Sun, Tears of the Moon
Copyright Syd Jones © 2011

Published by:
Signum Ops, 435 Nora Ave., Merritt Island, Florida 32952.

ISBN 978-1-4507-6523-7
Library of Congress Catalog-in-Publication data:
Jones, Syd
Atocha Treasure Adventures: Sweat of the Sun, Tears of the Moon

First Printed Edition, Release 3: April 18th, 2011

Graphics by:
Neal Sands

All photos by the author unless otherwise noted.

Manufactured in the United States of America. All rights reserved. No part of this book may be reproduced in any form or by any electronic or mechanical means including information storage and retrieval systems without permission in writing from the publisher, except by a reviewer, who may quote brief passages in a review.

Dedicated to "Unit 10", Captain Ted Miguel

Contents

- 1 .... Concentric Summers
- 9 .... Descent
- 14 .... Terra Agua
- 26 .... Wrecker's Wharf
- 41 .... Tropical Depression
- 50 .... Miguel's Navy
- 69 .... Company Man
- 86 .... Winter of Discontent
- 100 .... Looking Through an Ancient Mirror
- 109 .... Green Flash
- 116 .... Crescendo
- 121 .... La Margarita
- 136 .... Rogue Wave
- 154 .... Del Norte
- 169 .... 425 Caroline Street
- 187 .... Muddy Waters
- 201 .... Ebb Tide
- 222 .... Re Do
- 229 .... Lost and Found
- 246 .... Tribbits
- 265 .... Coda
- 282 .... May You Live in Interesting Times
- 295 .... Numero Only
- 310 .... Just Beyond the Horizon
- 317 .... Casa Key Haven
- 324 .... November 1997
- 331 .... Epilogue

FOREWORD

I've been asked many times over the years what was it like to find millions of dollars worth of gold and silver on the bottom of the ocean. The true (and short) answer is nothing comes easily from the sea. The real "treasure" of this fantasy activity turned out much differently than any of us expected.

This book gives the reader a unique, inside access to one of the most unusual companies that ever existed; Treasure Salvors Inc. It describes the company's quest to locate and salvage the richest Spanish galleons ever found, told from the perspective of an actual "front lines" participant. The events described were compiled from personal diaries and ship's logs spanning eighteen years.

By their nature, deepwater robotic vehicles or submersibles isolate their operators from their work. Treasure Salvors Inc.'s operations were more intimate and personal, human divers were the direct connection and often the deciding factor in the success or failure of the project. The numbers are impressive; the estimated value of the gold, silver, emeralds and artifacts that we eventually recovered on the 1622 fleet sites has exceeded 500 million dollars in value. Hundreds of man-hours were expended to search and locate each of the many hundreds of thousands of artifacts found. It was an extraordinary effort. The lack of regular pay, discomfort and exhaustion were only balanced by the pure wonder of being the first person in over three hundred years to touch an artifact from a lost Spanish galleon.

Not every employee that ever worked for Treasure Salvors Inc. is mentioned in the text, many quickly left the organization after discovering how hard and frustrating the work could be. Although many office personnel were not specifically named, I would like to acknowledge the dogged persistence of Sandy Dunn, Judy Sojouner-Gracier, Diane Jensen and Virginia Schultz in contributing to the company's success.

Second only to their unwavering optimism was Mel and Deo Fisher's ability to attract key people with just the right talents to their organization. Under the guidance of these few, the rest of us were encouraged to use our own skills, intellect, and determination to help make the dream a reality.

After seeing National Geographic film footage of us bringing up ancient gold and silver artifacts, I sometimes hear people saying: "I should have done that!" Well, perhaps... read on and decide if you really would have. Find your own treasure in life, whatever it may be.

Concentric Summers

Cayos del Marquis (Marquesas Keys) campsite

May 1626

As it approached the island, the longboat gently grounded while still some distance from the beach. In the clear, shallow water it was easy to see the sandy bottom's continued ascent towards the intended landing area. From here everyone would have to walk ashore and all supplies would have to be carried. Some sailors put away their oars as others jumped over the side into the warm, knee-deep water. In its lightened condition, the boat once again floated, and those in the water were able to push it closer to the shoreline until it firmly grounded again. The unloading could begin.

Still sitting in the stern of the boat, Francisco Nunez Melian looked back across the glassy ocean at the anchored mother ship to see if the other small boats leaving it were yet underway. Satisfied that they were indeed enroute, he slid over the gunwale into the shallows and joined the men from his boat in a splashy march to the beach. Melian wanted to be ashore before the other boats arrived so he could personally direct the initial unloading of equipment and provisions.

Everyone's morale had been high during the preparation period in Havana and the journey here. The men engaged in unloading Melian's boat were still obviously in good spirits. They talked and laughed energetically in the hot sun as they waded ashore with their heavy burdens. From the approaching boats the rhythmic sound of rowing and excited voices carried easily over the placid water. But when Melian himself reached dry sand, the reality of the island's barren isolation caused him a few private moments of despair.

There was no sign of the old encampment anywhere along the beach or near the barrier of mangrove tangles set further back from the shore. This particular island was just one of a group that formed a large, irregular circle protecting a shallow lagoon. Despite there being no indication now of the former camp's existence, he knew this to be the right spot. Some of the crew he had brought with him had been here with Vargas two years ago. Upon arrival they had all agreed that this was the island on which the camp had been. Even without their help, Melian would have picked this particular island as a base. It was the only one that was positioned well for the logistics of this venture. He instinctively knew Vargas would have come to the same conclusion.

Perhaps time, storms, or the Indians had eradicated any trace of the earlier habitation. It had originally been planned to quickly reestablish Vargas' old camp and get on with the real purpose of the expedition. Now it was quite apparent that they would have to start from the beginning, cutting a clearing in the mangroves on the higher ground away from the beach and establishing rudimentary living and defensive areas.

The rest of the small boats arrived and their occupants began unloading the cargo contained within. As the supplies and equipment began piling up on the beach, Melian showed some of the men where to start clearing an area out of the unruly foliage. One man suggested a quicker and easier method might be to burn out a clearing, but Melian decided against it. On a calm, clear day like today the smoke would be visible for miles. He preferred not to attract attention to their presence here, especially until the camp was established. Right now they needed all their energies focused on the expedition without having to worry about the intrusion of others.

After talking with his lieutenants about the camp's construction and the organization of the supplies, Melian agreed to a request that one of the small boats be allowed to reconnoiter some of the nearby islands in the group. All the ring islands were accessible from here by way of the central lagoon and there was a concern amongst his men about the presence of Indians on some of them. The men felt it was better to know now if there were Indians in the area rather than be caught off guard later. The conference slid into a debate regarding what to do if an Indian presence was detected. Becoming disinterested in further discussion of the subject, Melian dismissed them and their concerns and turned away. He walked alone down to the beach and along the water's edge. Let the others concern themselves with the trivialities of the camp. Now that they were here and the work had begun, he wanted only to concentrate his own thoughts on the real reason they had come. He had put too much effort into this whole venture to now be absorbed with minutia. As he stood facing the open ocean and the featureless western horizon, his task seemed bigger now than he had ever imagined.

It had been easy to speak confidently of success to the king's officials and other politicians back in Havana. Their shadow world of personal alliances and graft directed paths of political power and privileges all over the New World. Over the years it had become a system that Melian had learned to manipulate. Associations with the right people could ultimately prove very useful to someone with an agenda of self-promotion. For this singular project he had adeptly courted and cajoled the appropriate individuals to help his cause. At just the right moment in his campaign, Melian had announced a recent invention he had personally developed. He boasted that it would guarantee the accomplishment of his proposed quest. The momentum of his politicking and the announcement of this new device had culminated in the award of an exclusive Royal Charter from the Spanish hierarchy.

During his relentless massaging of egos in Havana, Melian had projected an absolute confidence in his ability to succeed in this venture. Others may have been more qualified for the undertakings but he had played a smarter political game. His bravado, however was a false front. The distilled truth was that he had gotten official title to undertake a type of work off these islands that he had never before actually done. Everything he valued was cast on his wager of success. If he were to fail at this, the hoped for riches, Royal favor and government appointment would be as far from his grasp as that of the man-of-war birds wheeling near the bases of the clouds overhead.

A sudden sharp pain on one of his exposed lower legs caused him to swat the skin there involuntarily. The biting fly was just another reminder that he was now removed from the comfort of civilization in Havana. It must have been doubly hard for Vargas to endure the discomforts of this place when his own labors had ended in complete failure. Melian remembered that the impatient Marquis de Caderaita had come here personally to spur on Vargas' efforts, but to no avail. Eventually they had all left with little more to show for their efforts than a place-name for these islands; Cayos del Marques. For an opportunist like Melian, their bad luck left a prize too tempting to resist.

As if on a flooding tide, his mood was buoyed upward by his own recovering optimism. Had not everything associated with this venture gone well so far? Having to rebuild the camps was just a minor annoyance. He was here in the right spot, poised with the men, supplies, equipment and good weather he needed. Tomorrow they would begin, and by all signs the end of this season would see his fortunes assured. Smiling to himself, he turned his gaze from the horizon and started walking back to the camp. As he trudged across the hot sandy beach, Melian remembered a conversation he had in Havana with a visitor from the Viceroyalty of Peru. The man had told him of a local Indian expression that described what he now sought. Sweat of the Sun, Tears of the Moon. It was an eloquent description.

August 1979

The narrow blacktop road roped over the rural Kettle Moraine countryside completely disposed to the deep blue-green of late summer. Central Wisconsin had been chiseled by ice ages and rounded by erosion into an ever-changing series of abrupt hills and deep valleys. Centuries of rain and freezing winters had smoothed most sharp edges from the landscape. The summer greenery further softened the geology's appearance. As the road approached a close set of steep-sided hillocks, it seemed to constrict in apprehension.

Rural roads like this are rarely used by anyone interested in the quickest journey to a destination. Direct routes offered by freeways or straight-line county road appeal to drivers who are more intent in their destination than the scenery. Traffic usually consisted of the very few residents or Sunday afternoon sightseers. Today it was empty. The late afternoon heat shimmered above the asphalt surface in places where gradually lengthening shadows had not yet reached. The hill's dense foliage sheltered hidden cicadas, conversing in an intense fugal buzz. Another tone, at first undistinguished from the insect's chorus, varied in pitch and increased in volume as the source grew closer.

The raspy exhaust note of the orange Lotus sports car rose insistently against the vehicle's own slipstream. It traced the tight arcs of macadam lightly, as a ballerina across a stage. The mature trees on either side of the road formed a gothic arch of green, with dappled patches of sunlight reaching the road's surface. At this velocity, under the steeply raked windshield, the alternate light and dark patches stuttered a natural strobe effect into the interior of the car.

I had previously driven this road many hundreds of times. Minute irregularities in the surface and the apex of every turn had long ago been committed to memory. Close friends whom I regularly visited lived 45 minutes from here. Half the pleasure of visiting them was the drive to get there. Motoring the empty country roads was, for me, a type of therapy. The simple narcotic of speed usually vaporized any of life's daily stresses by the time I reached my destination.

The road ahead remained clear. Normally, here I would be accelerating hard through the gears and concentrating. As the car reached the next ridge I did something I had never done before. I lifted my foot off the throttle, put the gear lever in neutral and let the Lotus coast to a halt. For a little while I sat with the motor idling, staring through the side window in thought. The car's interior soon became uncomfortable in the hot late-day sun. Turning off the ignition, I got out and walked to the other side of the road and then along its shoulder until I could no longer hear the popping and ticking of the cooling engine.

I sat on the edge of the road about 100 feet from the Lotus with my back to it. In front of me was a deep valley covered in field grass with more knobby wooded hills emerging on its western side. Sunset behind the hills was still a half hour away and there was no hint of a wind. The sun's disc faced me directly, and I felt the sweat flowing. The cicadas seemed to have raised their call's intensity with the closing of the day. I was viewing a completely benign pastoral scene, apparently a world in resolution.

Despite the tranquil atmosphere, my whole being felt off balance. It was a feeling that had been growing in me for some months. Now, even one of my favorite diversions, driving, wasn't enough to relieve the discord.

I had always been good at avoidance behavior. Activities that numbed the daily routine had often dominated my life rather than complimented it. Even during the frenzied scramble of obtaining a university degree, my life at school often became an afterthought to extracurricular interests. Despite being a good student and working part time jobs, my life's focus was always on unrelated diversions. The more I was forced to concentrate on serious responsibility the more demanding became my alternate world of avocations. Sports cars, skiing, flying, boating and scuba diving expeditions were amongst my favorites. Due to my lack of money, I could not always indulge in them at regular intervals. On the occasion when my limited funds allowed, exposure to any of these activities made up for the hours of drudgery expended on more mundane tasks. Until recently, this double life maintained my mental equilibrium. Now it was losing its effect.

I looked out at the hills as the sun's descent continued. In a few months this absolute green of summer's finale would be only a memory. With the loss of their leaves, the dark, brittle branches of winter trees would no longer hide the contours of the earth. They would be as insignificant as the few hairs remaining on a bald man's head. Wisconsin's bitter winters seemed to last forever.

Soon after graduation, I had begun teaching part-time and landed a job with a firm in Milwaukee. Settling into a regulated routine with a steady income, I now had more time and money to pursue my side interests. Despite the increased involvement with my hobbies, I experienced no satisfaction in a largely uninspired life. Why couldn't I simply work at something I enjoyed so much that side interests were unnecessary? Though my job was tolerable, climbing the corporate ladder for the rest of my working life seemed as appealing as enduring another Wisconsin winter. The problem was that it was hard to make a living at anything I really enjoyed doing. In picking potential vocations, fun-sounding activities didn't seem to make the list.

The lower portion of the sun's disc was just behind the upper branches of the trees on the western hills. I noticed the cicada's drone had subdued; the day was ending fast. When the rest of the sun finally sagged completely behind the distant hills, the landscape seemed to take a cool breath. It was now quite pleasant sitting here, and my thoughts drifted back to my dilemma.

Of all my interests, diving had been one of the longest-lived. I had read "The Silent World" by Jacques Cousteau at an early age and had never missed an episode of the 60's television series Sea Hunt. Exploring through a pile of National Geographic at my grandmother's house one afternoon, I found an article about a group of people who had discovered several Spanish shipwreck sites off the east coast of Florida. The pictures and descriptions of the recovery efforts were fascinating, and I read and reread the article many times. What a fantasy job to have! Though I could not picture myself ever actually doing something like that, I was inspired to learn more about the subject.

Through junior high and high school, I continued to read about ocean related topics. I read every book I could find on shipwrecks as fast as I could find them. After turning sixteen, I got my first scuba diving lessons during a family trip to Florida. Still, despite my obvious interest in the subject, future job prospects diving on shipwrecks were not promoted by the guidance counselors at my midwestern high school. In the late 60's, counselors put emphasis on more conventional careers.

Several years later after I had entered college, scuba diving and reading about shipwreck recovery projects still consumed my off time. I and other divers with the same interests joined a loose, amateur group, which tried to locate old shipwrecks in the Door County area of Lake Michigan. There was no profit motivation in our activity; we were driven solely by the adventure of discovery. Though most of our efforts failed to locate the sought after wrecks, there were a few successes. I can still recall the first shipwreck we discovered, the ominous dark image materialized from the gray-green as we swam towards it. The bitter cold of the water only seemed to heighten the suspense of the find. The dark, frigid, depths not only preserved the ship's remains but also shrouded the site with a somber, almost melancholy aura. Time had stopped for this vessel and everything it had carried. The discovery had been both a humbling and exhilarating experience.

In 1976, I happened to see a National Geographic special on TV. It told the story of a company endeavoring to find a Spanish treasure galleon called the *Atocha* and its valuable cargo, sunk somewhere off the Florida Keys in 1622. Despite a lot of intensive searching and some initial artifact finds, the whole expectation seemed doomed. Expensive court battles erupted with the State of Florida and the Federal government over ownership of the wreck. Several deaths had occurred when one of the salvage ships had sunk in the night. Two of those who died were family members of the people who owned the company.

I recognized some of the principals named in the story from other books and the original National Geographic article I had committed to memory at my grandmother's house. The story had since replayed many times on the public broadcast channels since the first time I saw it. The poignant ending

was still as powerful to me as the first viewing had been. It implied that despite its many problems, Mel Fisher's little company would somehow continue in its efforts to find the *Atocha*. I heard no news about the firm's subsequent fate since the National Geographic special first aired in 1976.

From my perch on the side of the road I looked back at the Lotus. The orange paint was still vivid even in the dimming evening light. Suddenly restless, I got up and started to walk back to the car while thinking about the two women I had been dating. One I had been seeing for about a year and the other off and on for nearly six. Both relationships had reached their high water mark; it was obvious that neither was really going anywhere. Like all my diversions, they were just condiments to the main course of life; sometimes spicy, but not very filling.

Before I got into the car I took one last look across the valley. Still visible in the twilight were hundreds of fieldstones among the grass. Rounded by the action of ice age glaciers that tumbled them across this region, they lay where they settled when the ice finally melted. Those simple stones spoke volumes to those who understood their origins. In their patient dignity, they told of extinct worlds and futures yet to be. Though I would not know it that night, in one month I would be enroute to Key West, Florida, leaving behind the Lotus and all it represented.

DESCENT

Two days of driving from Wisconsin brought me to central Florida. My mom had accompanied me this far and I dropped her off at her sister's house in Melbourne. Since she rarely got to see her relatives, my trip south was a perfect excuse for an improptu visit. At the end of her stay, she would catch a plane back home. I continued alone to Key West in my well-worn Fiat sedan.

Just a little over two weeks ago, a simple lapse in the rhythm of work, triggered events that now led to my journey down this highway. Had I not already been mentally primed for a change in life's direction, the occurrence would have had no effect. I had been sitting at my desk in front of a computer terminal, absorbed in melismatic office routine. Information that I had requested from the computer was slow in coming, and the long wait broke the cadence of the task. My concentration faded, allowing me some daydreaming time while the machine continued to search. For no particular reason, the ending of the National Geographic documentary on the *Atocha* expedition once again flashed in my head.

Though it had been some time since I had last seen the film, I still vaguely remembered the name of the company involved in the project. The computer terminal beeped, indicating that it had finished its task. The sound interrupted my daydream, but not an end to my thoughts on the subject. I picked up the phone and dialed directory assistance in Key West. The operator found a listing for a business under the name I remembered. I wrote the phone number down on my desk calendar.

On a whim, I automatically dialed the number. As the phone rang on the other end, my reserves of courage began to collapse. Before I could hang up, a pleasant sounding female voice answered "Treasure Salvors." My reply was automatic and unplanned. "Uh, Hi, My name is Syd Jones and I've been following your company's exploits over the years. I am a certified diver with wreck diving experience and I've spent a great deal of time on boats of various types. I know how to fix engines and other mechanical systems. I'm coming to Florida on an extended vacation and I was wondering if you might be interested in some volunteer help on the *Atocha* project. I think I have enough money to stay anywhere from six months to a year."

I stopped talking at that point, amazed at the content of my improvisation. The lady on the other end responded in a measured but courteous fashion as if she had already fielded dozens of calls of this type; "Well, we don't take on volunteer divers but if you are ever down in Key West you may want to contact our operations manager. Perhaps you may have some skills in which he might be interested." She gave me the company's address and the name of the operations manager whose name sounded like Ted McGill. As quickly as it had begun, the phone call was over.

I sat back in my chair and thought about what I had just said. The details about the diving, boats and mechanical skills were true, but I really hadn't thought about being a volunteer. Perhaps my subconscious felt too presumptuous to ask for a paying position due to my lack of professional experience in the field.

As far as the "extended vacation", my hobbies and other interests had effectively consumed all of my excess money. I would be lucky if the eight hundred dollars nervously residing in my bank account would last six weeks, much less six months. Still, the lady on the phone hadn't said they weren't interested, just that I needed to talk to the guy who did the hiring. Though in reality her response to my inquiry was certainly more nebulous than positive, I immediately felt a mood shift. Suddenly, everything seemed possible.

Under the current date on my desk calendar I wrote GIVE NOTICE then paged ahead two weeks and marked LAST DAY. Rising from my chair, I went over to the desk of a co-worker who had become a good friend. Like me he was a secondary cog; at this job we were just two of the many unnoticed small wheels whirling in the machinery of business. During breaks and off-hours, we often talked about climbing the corporate ladder or perhaps shunting the whole executive process by somehow discovering some mystical life experience with more meaning. We were both fresh out of school and were dissatisfied to find ourselves in the quagmire at the bottom of the corporate world.

Leaning on his desk I announced in a rather self satisfied manner; "John, I'm quitting this place and moving to Key West to dive on shipwrecks for a living." It was apparent from his facial expression that this announcement didn't contain the shock value I had hoped for. He probably thought I was kidding him about some imaginary lifestyle. We sometimes liked to joke about such things over a beer at the end of the day. To punctuate my statement, I announced, "I'm going right now to the boss's office to give two weeks notice." And I did.

Later that afternoon I wrote a letter to Ted McGill, explained my conversation with the lady at Treasure Salvors and my intentions of leaving for Key West in two weeks time. Enclosed was a brief resume, tilted as appropriately as possible. The momentum carried by my decision to leave had already pushed aside any reservations I might have had earlier. My only concerns were to attend to a few personal matters and decide on what possessions I should take and which I should store.

The sudden decision to move to a place that I had never previously been was not as rash as it sounded. Though no job there had really been offered to me, I was confident in my ability to find some type of employment in Florida if this didn't work out as planned. With no wife, kids, mortgage or other serious responsibilities, there was no pressing financial reason to bind me here. Youth was also a factor in the equation. I was still young enough to be flippant about the idea of settling down. The idea of a new direction in my life was intoxicating.

Not everyone shared my enthusiasm for this change. My father in particular was a hard sell. As a long time businessman, he viewed my present employment as a valuable stepping-stone. An early career mistake might negatively influence my future. This became a common topic of discussion over the next few days. His resistance to my plans only increased when I received a letter from Treasure Salvors in response to my resume. It basically stated that if I had the qualifications they wanted, that they would pay me their normal rate of fifty dollars a week. The letter also stated that there was a great deal more work involved rather than just diving, and that most of the divers lived aboard the salvage boats. Food was supplied for the crew while out at sea but not in port.

My mental picture of a salvage boat was one of a crude, rusty barge-like vessel. The thought of living aboard something like that wasn't my first choice of accommodations, but at least they were free. My father never got to that point in the letter. He stopped at fifty dollars a week. His attitude about the situation was that I was throwing away a good paying job for the experience of living like a wharf roach.

At the age of twenty-seven I certainly didn't need my parent's approval to leave my job or Wisconsin. In deference to family harmony, I preferred to depart with them feeling at ease. The type of work I was hoping to get might not offer much financially, but it did imply a certain degree of adventure. Perhaps the job satisfaction that eluded me might be easier to find at Treasure Salvors. I needed to assure them that no permanent harm would come to my future by deviating off-track for a while. Psychologically, it would be much easier on all of us if I left with them in agreement with my motives.

Fortunately, my mom was enthusiastic about my plans. She had traveled extensively at the close of World War Two. In recounting her own experiences, she helped ease my father's negative attitude. As my departure date approached, his objections reduced to a conditional acceptance.

The situation with the two women whom I had been dating presented another problem to resolve. Lorrie had been a long time companion, but our relationship was well past its zenith. Joan had been a more regular companion over the last year. She was an energetic, outdoorsy type with an easygoing personality. My new plans didn't include either of them, but I needed to explain my intentions. Aside from any romantic aspects, I had come to value their respective friendships. Both women knew vaguely of each other's presence in my life, but neither had pressed for details. My lack of potential for permanent commitment was apparently understood by all of us.

When I told Lorrie of my impending departure for Key West, her reaction was one of haughty disbelief. She parroted my father's concerns and confidently predicted my quick return to Wisconsin. Joan reacted much differently. She enthused about the possibility of diving on a professional expedition. Any bad feelings she may have had about me leaving seemed masked by genuine excitement.

During the last year, I had started the expensive process of getting my pilot's license. If I secured a job at Treasure Salvors, the salary of fifty dollars a week would stall my flight training for the time being. Having recently soloed, my urge to fly one last time before leaving was hard to resist. Since Joan had sometimes accompanied me on earlier trips to the airport, she offered to drive me there for a last solo flight on the eve of my departure.

After my last day at work, I began the organization of my belongings. I threw out some, but most of my stuff ended up in boxes for storage at my parent's house. I planned to take only my dive equipment, camera and a few clothes. If I were to get the job in Key West, there would likely be space restrictions living on a boat. For the excursion south I drafted the old, anonymous looking Fiat that had served me as my daily commuter car.

The evening before my departure was clear, cool and still. Even as I busily stored the last of my possessions in my parent's garage, I found it hard to ignore the atmospheric calm. All pilots, especially fledgling ones, live for such conditions. If this was to be my last flight for a while, at least the weather was sympathetic. Just as I finished stacking the last of the boxes, Joan arrived to take me to the airport.

The ride to the airfield was numbed by a counterpoint of small talk.

Joan commented about the weather and reported things she had done at work. I kept waiting for her to bring up the subject of my leaving, but she carefully skirted it. Maybe avoidance behavior over potentially difficult subjects was something we both had in common.

After my flight, the ride home was almost a carbon copy of the trip to the airport. When Joan dropped me off at my parent's driveway, our good bye was as casual as if we were seeing each other later that night. Only as she drove away could I see sadness in her face. As her car disappeared down the street, I was drawn in from the twilight to the lights in the still open garage. In boxes stacked along one inner wall were all my possessions with the Lotus as a centerpiece parked amongst them. Until recently this pile of things had represented my life. Contained within the boxes was the detritus of my various interests. The events, people and emotions associated with them now seemed out of focus and distant. The boxes looked all the same with little to differentiate the contents of one from the other.

Was I really making the right move? I had systematically said goodbye to my friends, romantic ties and the avocations that I had spent so much time pursuing. Most people would have been content to have had the opportunities I had already enjoyed. Why throw all this to the wind just on the remote possibility of doing something I had really only seen on television?

This last minute shudder of self-doubt might have halted my journey south before it started, but I had yet to experience the creeping conservatism that comes with age. Rather than dwelling on the comforts of my present situation, I was fixated on the promise of a grand new adventure. In my young mind the worst that could happen would be that I got no job at Treasure Salvors. Certainly I could find other jobs in Florida. What money I had would last until I found employment of some type. At least I could ride out the coming winter in comfort rather than freezing with the rest of Wisconsin.

The arrow straight Florida turnpike was set in a flat expanse of scrub and farmland. During the trip south the Fiat's engine had been droning at the same pitch for hours. The heat, sameness of scenery and traffic were conducive to reflection about the last few weeks. Only when the highway neared Ft. Lauderdale and Miami did the highway's congestion require more concentration. At Homestead the turnpike ended, and the long single road that led to and connected the Keys began. In driving through Key Largo and descending through the island chain, my rising expectations pushed aside thoughts of the recent past. I had dealt this hand of cards, and I would soon see how it played out.

TERRA AGUA

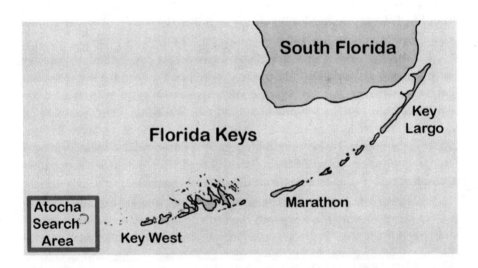

Most motorists who visit Florida never make it to the Keys. The mileage from the Georgia-Florida border to the islands is the biggest obstacle, and most attractions usually visited by tourists are found in other parts of the state. Those who ventured south of Homestead might be surprised at the lack of traditional Florida vacation points of interest. There are few real beach areas, and many of the islands are barely more than stands of mangroves straining to stay above high tide. From the air they appear as flat, irregular green stepping stones leading from the Florida peninsula into the Gulf of Mexico.

Key West is the last inhabited island in the Keys chain. A single highway connects it and the other populated islands to the mainland. The road traverses the watery gaps between the islands via an exceptional system of bridges. New, relatively wide and modern bridges were completed in the mid-1980's, making driving through the Keys much less stressful. The original older, narrow road that defiantly perched on bridge foundations originally intended to support a single railroad track had finally been replaced.

My first passage down through the Keys was in the old bridge era, when the vista along the way was much different then than now. Due to the close proximity to the mainland and the excellent diving conditions, the first island, Key Largo, was a major tourist destination by Key's standards. It sported a few small dive shops and even a modest Holiday Inn. These businesses capitalized on Scuba divers and snorkelers interested in visiting John Pennekamp Underwater State Park, which is on a portion of the coral reef tract just offshore.

Aside from diving oriented establishments, there was precious little development visible from the main road. In the future, blocks of condos, strip malls and fast food restaurants would homogenize Key Largo and spread along the entirety of the Keys. In 1979 however, Key Largo's level of development beyond its sister islands wasn't much above a rural crossroads when compared with other Florida tourist destinations.

As the road continued southwest from Key Largo, the occupational nature of the local inhabitants began to change. There were occasional Mom and Pop gift shops and small motels lurking hopefully in clearings next to the highway, but commercial fishing boats and related industries were more prominent. The weathering effects of salt air and harsh sun combined with a generally casual attitude about appearances made it difficult to date many of the structures and vessels seen along the way. Tourists enthralled with the manicured, antiseptic presentation of Disney World in Orlando would have found the Keys to be a somewhat distressed and tacky place.

Further along toward Key West, one more barrier against the weak-willed remained. In the upper Keys, the distances between the islands and their connecting bridge spans were fairly short. Approaching Marathon, about halfway down the chain, the gaps and their bridges start to stretch out. The highway bridges were built on the old railroad structure that businessman Henry Flagler originally constructed at the turn of the century. A powerful hurricane destroyed the railway in 1935, and the surviving rail bed and bridges were eventually converted into a highway that tied the Keys to the mainland.

Since the original bridges had been built to support a structure the width of a railroad track, the highway added on top had to also be quite narrow. The two opposing lanes constructed were barely wide enough to let normally sized vehicles pass. There was virtually no shoulder in case of emergencies or breakdowns. On the immediate outside edge of the road were sections of old railroad track welded together to form a rail to guard against vehicles tumbling off into the ocean below.

A portion of the original Overseas Highway near Bahia Honda

Driving on these bridges was a white knuckle affair at the posted 50 mph speed limit while trying to avoid the unyielding guard rail inches to the right and oncoming traffic weaving in the cross winds. Those with a real taste for adventure could make the same trip on these long, unlighted causeways at night.

Most people would imagine that traveling through these small, flat islands would give them an uninterrupted view of the shoreline and the sea. Actually, many of the Keys had a fair amount of dense vegetation along the roadside, and where the highway ran along the ground it was difficult to see the water. Only when driving out across the long bridges could the magnitude of the ocean surrounding these isolated dabs of land be understood.

Those of us who live in cities or suburbs far inland often give the natural world the same regard as we would give a domesticated animal. Our lives are spent in climate-controlled houses, cars and jobs. Wilderness is represented by a wooded park or farmland. The weather, though uncontrollable, is rarely more than a brief inconvenience during the few steps from the shelter of a parked car to a building. We see nature like a cow or pet dog; some vestiges of a distant wilder past still remain, but overall human influence has tamed out any real threat to our well being.

A different attitude prevails in the Keys. With land elevations barely above sea level and a location completely surrounded by open ocean, nature is THE dominant force. On some Keys, the long construction boom beginning in the 1980's produced expansive resorts, hotels, and private dwellings in densities similar to a mainland city. Despite the increased presence of concrete and steel, man's structures here simply don't imply our perceived dominion over the natural world as much as in other places.

The Keys show their innate wildness in many forms. Along the island chain there are impassable mangrove thickets on the hundreds of uninhabited islets, which provide shelter for birds and sea life. Between the Keys and the Gulf Stream lies an exotic gallery of coral reefs, lying just below the water where many ships have died. There are vast, deep-water expanses sometimes driven by violent hurricanes. As terrestrial animals, humankind's most basic need is dry land. At nature's whim, dry land on the Keys can be in very short supply.

After passing through Marathon and negotiating the famous Seven Mile Bridge, I could see that the lower Keys were even less populated than the upper ones. Although I had visited Key Largo previously, my knowledge of what lay beyond was purely based on my imagination. Like most people, my perception of tropical islands was of palm shaded, pure white beaches massaged by foam-laced indigo surf. The "Lower Keys" had no beaches and few palm trees; the shallow, green tinted water around them was as flat and unromantic as the islands themselves seemed. It slowly became apparent that the reality of Key West might be much different than my previous notion of it.

About one hour's drive past Marathon the traffic noticeably increased along with the number of buildings alongside the road. Suddenly I drove past a sign that welcomed me to Key West. Having no idea where Treasure Salvors was located, I stopped at a gas station to ask directions. The attendant told me to continue driving through "new town" until I got to a bridge on the right that would take me to "old town". A right turn on Simonton and another right turn on Front St. would lead me to "the old ship place".

Pulling out onto the road again, I continued as directed. The area the gas station attendant described as "new town" was an unremarkable procession of 1950's and 60's vintage buildings, housing shopping centers, restaurants and other everyday businesses. With the exception of Florida Bay, which lapped against the right side of the roadway, Key West had the look of any generic small southern town. In a short distance was the Garrison Bight Bridge. I crossed over it and soon entered what I guessed was "old town". Though my attention was focused on watching for street signs, I noticed that this area was primarily residential with older wooden houses built very closely together. Hardly any cooling breeze came through the car's open windows in the slow moving traffic. The rising temperature in the car matched the increase of nervous energy inside me.

As I turned onto Front Street, I spotted Treasure Salvors' location. In the National Geographic documentary about the company, the firm's unique headquarters had been featured several times. It was a life-sized replica of a Spanish galleon tied up to a dock in the harbor. In the film, the galleon had looked glorious, and apparently housed a museum as well as the company's offices. It was hard for me to imagine a more appropriate structure to symbolize the group's function.

A tour guide waits for visitors aboard the *Galleon*.

The Fiat squeaked into the potholed, gravel lot that served as a parking area adjacent to the *Galleon*. I had arrived at last. The sudden reality of actually being here made my throat a little tight. In Wisconsin I had conned myself into believing that there was a good chance of my somehow getting hired by Treasure Salvors. The self-assuredness that had been guiding my direction to this point dissipated with my arrival. Before my confidence wilted any further, I parked the car and took out a copy of my resume from a folder on the passenger seat.

As I walked toward the *Galleon*, I saw that hard times had fallen on the ship in the years since National Geographic had filmed it. The bowsprit and forward mast were completely missing, and what little remained of the rigging dangled uselessly. It looked more like a shipwreck itself rather than the headquarters of a famous company. A short, wide gangplank served as the entrance to the vessel, and as I crossed over it I passed several scruffy looking individuals mixing cement in a wheelbarrow.

The gangplank lead to the main deck area. There was a large bronze cannon sitting on its carriage in the center of the deck with a haphazard array of well used marine equipment piled along the far side. Facing the back of the ship, I noticed an opened arched door leading into the sterncastle. Movement and voices coming from inside compelled me to enter.

Inside was a large, dark wooden room with several display cases and a U shaped counter along the forward facing wall. There were steps leading downward in the middle of the room and at the back wall was a closed door. A few people I presumed were tourists were looking at the contents of the various display cases. A woman, self-absorbed in a magazine, stood standing behind the counter. With the most self-assured voice I could muster, I approached her and said, "Hi, My name is Syd Jones. I'm here to apply for a job… I talked to a lady here on the phone about it. I sent Ted McGill my resume and I believe he's expecting me." She smiled back in a half-interested manner. "Well I guess you need to fill out one of these." She reached under the counter and pulled out a folder marked "job apps", opened it and handed me a blank standard job application form. Before turning back to her magazine she said; "I don't know when Ted's coming in. I haven't seen him all day but he's usually here by now. Why don't you sit down and fill that out and he'll probably get here before too long."

Against the railing next to the downward stairway was a crude looking bench with an object sitting on one end. With no chairs in sight, the bench looked like the only place to sit. I walked over and sat down on the unobstructed end of the bench, intent on filling out the form before Ted McGill arrived.

When more tourists wandered into the sterncastle the woman at the counter answered their questions regarding tours of the museum. She explained that the exhibits were located in the forward and lower portion of the ship and that the admission price included a personally guided tour. Groups were taken through on an hourly basis and since it was nearly 3 o'clock, the next tour was about to begin. Since I was sitting only a few feet away from the counter, it was hard for me not to be distracted from the boring task of filling out the job application form. I paused for a moment and watched while the woman expertly herded the tourists out onto the main deck area for the beginning of the tour. Once outside the door opening they were beyond my field of vision, but I could hear the woman speak in an elevated volume;

"Welcome aboard the *Golden Doubloon*. This ship is an exact replica of a 17th century Spanish galleon used to transport millions of dollars worth of gold, silver, and other valuables back to the Old World from the colonies in the New World. These ships not only carried rich cargoes but also served as transportation across the Atlantic Ocean much like jet airliners do today. There were dangers along these trade routes. Pirates and, more often, hurricanes claimed ships, people and much of Spain's wealth."

"Our company has found several of these types of vessels which were shipwrecked in 1715 along the east coast of Florida and is now concentrating on locating two more ships that sank near Key West in 1622. You may have heard about one, the *Atocha*, from the National Geographic movie and magazine article about our company. We have located a small portion of that shipwreck and are still looking for the rest of it along with another galleon, the *Santa Margarita*, which sank in the same storm. The *Atocha* is reported to be one of the richest ships ever lost. Follow me now into the forecastle, and I will describe the artifacts we have on display."

I heard the noise of diminishing footsteps as the crowd obediently followed the tour guide into the front of the ship. With nothing else now to hold my attention, and the job application completed, I took time to look around the room more carefully.

On the same bench just a few feet away from where I was sitting was a lump of something that looked about the size of a loaf of bread. In my haste to complete the application form, I hadn't paid attention to the object. Now on closer inspection it became obvious that it was a silver ingot. It was slightly dark from tarnish and rough on all sides but the top. Imprinted in the smooth upper surface were crude looking monograms and Roman numerals surrounding a small divot in the center. Next to the bar lay a small white cardboard sign that read; "ATOCHA SILVER BAR - VALUED AT $50,000" The last "O" and the comma on the price had been added on with a different color pen, so it was difficult to tell if the amount was a joke or legitimate. Either way it was a valuable artifact and I was surprised that it was sitting out in the open, completely unguarded. Perhaps it was a replica, put on this bench so tourists could get their pictures taken sitting next to it.

I reached over and touched the impressions stamped on the top and then tried to rock it, but it resisted my effort as if it was bolted down. As I stood up, I grabbed the bar with both hands and lifted. The effort it took for me to raise the ingot off the bench surprised me considering it' compact dimensions. I gently set the 85 lb. bar of silver back where it originally rested.

Little wonder that Treasure Salvors felt secure leaving it out where people could touch it. A thief wouldn't be able to get far on foot trying to carry off a load like that. Still, the weight and the monetary value of the bar didn't interest me as much as its historical presence at which the markings hinted. Sitting next to it almost seemed like an honor. Looking around the room's display cases proved somewhat anticlimactic after the silver bar. Most of the artifacts on display in this room were things I'd encountered before in other museums. Perhaps the most interesting items were kept in the main exhibit below decks.

The U-shaped counter offered the usual tourist accouterments for sale. Among the Key West souvenirs there were T-shirts with "ATOCHA EXPEDITION 1979" lettered on them. If I couldn't find employment at Treasure Salvors, at least I could by a shirt proving I had been here.

After I had thoroughly explored the contents of the room, my wait for Ted McGill began to get tiresome. The upbeat mood that had accompanied me here was being replaced with an attitude one normally experienced while in the waiting room of a dentist office. Fortunately the voices, which emanated from the stairwell in the center of the room, were a refreshing diversion. The tour group returned from the lower areas of the museum back up to the sterncastle area via the stairway. They had been gone about 45 minutes.

Most of the people in the group left immediately as soon as the guide thanked them for their patronage. A few lingered around the counter to examine the contents of a box that the tour guide quickly extracted from a small safe sitting beside the wall. As I heard her talk about Spanish silver coins, I moved closer to the counter to get a better look. Before I could see what she showed the tourists, her concentration was broken when she caught a glimpse of a man who walked in through the doorway. He nodded at her as he strode past the counter in the direction of the closed door at the back of the room. I watched him pass as she said; "Ted, there is someone here to see you." As he turned, I stepped toward him with my hand extended.

"Hi, I'm Syd Jones. I sent you a resume several weeks back and came to Key West to talk to you about job possibilities. Hopefully you got my letter telling you I would arrive today." He shook my hand with an expressionless smile on his face and paused before saying "Oh yeah, weren't you from Michigan or one of those other cold places?"

"Wisconsin. You know, the dairy state."

"Up there milking' cows eh? I need to get some coffee…follow me and we'll talk back here." He turned and walked toward the door at the back of the room.

I followed his lead through the doorway into a narrow hall with small rooms off to the side. We entered a room on the left that was little larger than a broom closet. There were coils of electrical cable and equipment stacked on shelves along the wall. A stained coffee maker and cups perched on a small table to one side. He poured some coffee out of the pot and began to add some cream. "Want some?"

"No thanks, I never really developed a taste for it."

After briefly stirring the contents of his cup, he took a quick swallow. Apparently satisfied with the taste, he reached for the papers in my hand. "Well, let's see watcha got here."

He spread my resume and job application form out on the table next to the coffee maker. His demeanor while he looked over the papers was similar to that of a teacher grading a difficult test. Ted deviated from reading only to take the occasional sip of coffee, and seemed to ignore the fact that I was standing right next to him.

His slightly frosty manner and the way he concentrated just a little too hard on the paperwork put me off. In most successful job interviews I had in the past, the person doing the hiring casually skimmed the resume and then asked lots of questions. Ted said nothing at all, and in the uncomfortable silence I had the opportunity to study him. He was a little shorter than me, maybe around 5'11" and of moderate build. His black hair was short on the sides and the longer portion on the top was combed straight back. The earpieces of his black-rimmed institutional looking glasses highlighted his gray hair's progression through longish sideburns. A faded plaid shirt hungout over drooping khaki shorts and his scuffed work boots were an indication that dress codes here were a little less important than at my old office.

"What else can you do?" He had finished reading and was suddenly looking at me straight in the eye, waiting for my answer. His broad question caught me off guard. I had massaged the information on the resume to be as complimentary as possible to the type done by work Treasure Salvors, so there wasn't much more I could add.

"Well, as you can see on the resume, I've spent a great deal of time around all kinds of boats. I know boat handling, maintenance, and navigation. I know how to do rigging and splicing."

"What else?" He was looking at me with the same emotional display on his face as an Easter Island statue.

"I've raced automobiles and am good mechanically. I know how to rebuild engines and troubleshoot electrical systems." As the details of my educational and diving background were covered in the resume that he had so carefully read, I wasn't quite sure what to tell him without specific questions from him. Maybe it was time to improvise with generalities.

"I'm the kind of person who is very goal oriented.... I like to finish what I start. The type of work you do is something I've been interested in for some time. I've done a lot of study on old shipwrecks and I think with my enthusiasm for the subject I would need a minimum amount of direction to complete a task."

"What else?"

There was still no emotion discernible in his facial expression to hint at the direction the interview was going. I answered his question by elaborating on details from my resume. Since my reply was primarily a repeat of what he had already read on paper, I hoped he realized that I was running out of material.

"What else?"

By now I had told him everything but my great grandmother's blood type. If this interview had been a boxing match, I would certainly be on the ropes. Though incongruous with the position for which I was applying, I, in desperation, used my past job's experience as a reply.

"What else?"

It seemed that he was fishing for an answer that I was too ignorant to give. Perhaps his interview style was a quick and humbling way of showing me that I had no qualifications of interest to him. It was time for me to regroup and try to bow out as gracefully as possible. Instead of trying to answer this last question, I would apologize for wasting his time and leave.

"Ah, Mr. McGill..." He started talking over me before I could finish my sentence.

"My name isn't McGill - it's Miguel. I'm always being confused with a bloody Irishman. What did they tell you we paid divers here anyway?"

"Fifty dollars a week."

"We TRY to pay fifty dollars a week... sometimes you may have to be happy with less. We don't have a lot of money right now so payroll can sometimes be a little... irregular. You'll eventually get any money owed to you, it's just that we can't always pay the crews every Friday and still have money left for the operation. Can you handle that?" I could handle anything except another round of "What else".

"Uh, sure, that's no problem."

"OK, then I want you to go over to the *Swordfish* - it's the black and gray boat tied up next to the *Galleon*. Tell the crew there to find you a bunk. It's too late now to talk to the bookkeeper. Tomorrow morning go upstairs to his office and tell him I put you on the payroll. He'll have some forms for you to fill out."

Was I hearing this right? It sounded like I had just been hired. "Great... by the way, what time does work start in the mornings"

He finished off the last of his coffee and wiped out the cup with a paper towel. "We start at 9:30 while we're working in port. As soon as you get your paperwork with the bookkeeper taken care of, we'll find something for you to do. Right now I have some other business to deal with, so I'll see you tomorrow." With the conclusion of that statement he turned and walked out of the room and disappeared further down the hallway.

"Baroque Dolphin" cannon handle

I retraced my steps back into the sterncastle display room and then through the arched door into the sunlight on the main deck. Despite all of the personal and emotional obstacles along the way, my hopeful daydream had become a reality. During the course of the interview, Ted hadn't given me any indication of what I was being hired to do, but at least he had granted my admittance to the company.

Before going to the port side of the *Galleon* where the salvage boats were tied up, I paused for a more careful look at the bronze cannon I had previously noticed. It was one of the relatively few items actually recovered so far from the *Atocha*. Its size and mass were accented by the green patina of the metal. On the top center portion of the barrel were two leaping shapes known as Baroque Dolphins. Though only a stylistic caricature of a real dolphin, I had read that the design was popular in the 17th century for superstitious reasons. Sailors believed the presence of dolphins brought good luck, and would sometimes rub a cannon's dolphin handles before a long voyage or sea battle to promote a favorable outcome. I reached out and rubbed my hand over the dolphins. It was not so much for the future, but as a confirmation of my luck.

 Wrecker's Wharf

Portions of the two salvage boats tied alongside the *Galleon* were just visible above the main deck railing from where the cannon stood. I walked over to the railing for a better look. The nearest vessel was painted black with a gray superstructure and looked as if it was about 60 feet in length. It had lean, purposeful lines normally associated with military patrol ships and was nothing like the rusty, barge-like craft I expected would be here. On the side of the bow was painted the profile of a leaping swordfish. From the radar mast just behind the wheelhouse to a framework that supported a canopy above the back deck ran several small ropes that were serving as clotheslines. Sheets, towels, and clothes were spread along these lines, which visually suggested the combination of a yacht's flying pennants and a squatter's camp. The roof of the main cabin between the wheelhouse and back deck was piled with plastic gas cans, life rafts, outboard motors and coils of rope.

Tied next to the *Swordfish* was a smaller, beamier and obviously much older boat. It had the looks of a tired wooden commercial fishing boat from the 40's or 50's and was painted a very faded blue. Both boats had large elbow-shaped tubes mounted just above their sterns, a feature I recognized from the National Geographic documentary. They were called propwash deflectors, blowers or "mailboxes", and were used for excavating underwater. When lowered into the water, one end of the tube fit over the ship's propellers and the other end pointed straight down. When the engines were put into gear, the propellers forced water through the tubes and downward toward the ocean bottom. The force of the water would push away the sandy overburden, exposing any artifacts that might lay buried there.

Treasure Salvors Inc.'s 1979 salvage fleet, *Swordfish* (left) and *Virgalona*
Photo: Leah Miguel

I climbed over the *Galleon's* wooden railing and onto the black and gray boat and followed the catwalk aft in the direction of voices and movement on the back deck. The damp laundry lazily flapped across my path in the hot breeze as I pushed it aside like an explorer through a thick jungle. As I passed one final soggy towel, I arrived under the canopy-shaded area of the back deck.

Next to the railing opposite from where I stood was someone playing a guitar and sitting on what looked like an old bus seat. In the middle of the deck a stocky, disheveled individual rooted through a wooden toolbox. My presence caused the guitar player to stop in the middle of his song. He leaned slightly forward and asked, "Can I help you?" His voice had a southern lilt. "Hi, Ted just hired me and told me to come over and find a bunk on the *Swordfish*. I'm Syd Jones."

The guitar player was of medium build, had thinning dark hair and a mustache. He smiled and raised his hand toward me. "I'm Fred, and that's Joe." I shook Fred's hand and then turned towards Joe. He looked up from his toolbox, extended a greasy hand with an unenthusiastic, "Hey". Fred pointed at doorways that lead downward from the back deck into the cabin area. "Go down below and see Wally the cook. He'll show you where there's an open bunk. Where's all your clothes and gear?"

"I left it all out in the car. I didn't want to bring it with me until I knew where I was going."

Fred and Joe looked at each other like they had just found a 20 dollar bill on the street. "You've got a car?" sounded in unison.

"Yeah, an old Fiat."

"Would ya mind giving me a lift over to the laundromat later? It's pretty far to walk with a load of laundry." Fred looked at me with all the puppy dog innocence he could conjure up.

"Sure, no problem. Let me get my stuff on board and then we'll go." He nodded with approval and I proceeded down the stairwell and into the cabin.

The main cabin had a narrow central walkway bordered by two large elevated bunks and a small dining area at the forward end. Beyond was a doorway with a short set of steps leading downward into another cabin area beneath the wheelhouse. As I approached this last set of steps, the smell and sounds of someone cooking nearby coalesced in the narrow passageway. Upon entering the lower cabin, I saw the doorway to the head and shower on the left, another bunk area directly ahead and a partition on the right. Stepping past the partition brought me into the galley and the source of the aroma. A person I assumed was Wally was armed with a spatula, busily flipping hamburger patties in a well-worn Teflon frying pan. He glanced at me briefly before his attention was caught by the appearance of a large palmetto bug crawling up the bulkhead from behind the stove. Wally tunelessly chanted the opening phrase of "La Cucaracha" and lunged with his spatula. The remains of the Palmetto bug drooled down the wall looking like a blob of used chewing tobacco with feelers. Wally cackled to himself as he went back to checking the burgers without wiping off the spatula.

"You must be Wally" I said. "I'm Syd Jones - I was just hired today. The guys on the back deck said for you to show me an empty bunk space down here."

He turned off the stove and began to pile the burgers on a greasy paper plate. "Nice to meet cha!" We shook hands. "You hungry? There's plenty of food."

My mind's eye superimposed the crushed Palmetto bug onto the hamburger patties. "No thanks, I had a late lunch."

Wally lit up a cigarette, which quickly filled the small compartment with smoke despite the open hatch above our heads. He pointed to a bunk just visible under a pile of old military style duffel bags, stained pillows and dive gear. "This one here is open. Everybody's been leaving their shit on it. Most of this junk needs to be thrown out anyway." With that he started grabbing items on the bunk and tossing them onto other bunks or off to the side. "There you go; home sweet home. You're gonna love it here, every day is Sunday!"

It was early evening by the time I had unloaded the car and stowed my clothes and gear aboard the *Swordfish*. As soon as I finished, I found Fred and told him I was ready to take him to the laundromat. We got into the Fiat, and he guided me through the still hot, musty smelling streets of old town. Our conversation during the brief trip was mostly a polite exchange about where we both were from, though Fred's real interest in me seemed based on my car ownership. At fairly regular intervals our discussion drifted toward other destinations to which he would like to be driven in the near future. As a footnote he also informed me that he was the first mate on board the *Swordfish* and was in charge of the crew when Ted wasn't around.

When we returned to the *Swordfish*, I went back below decks to finalize the organization of my new living area. In the passageway I ran into a few other crewmembers I hadn't previously met who introduced themselves using only their first names. I had already noticed that everyone I had met so far had never offered his or her last names. Over time I learned that Key West was a first name only kind of town…for many reasons.

Later while arranging my things, I heard the sound of approaching footsteps in the passageway preceded by a friendly sounding voice. "Hey, you the new guy? I just got back from working on Mel's houseboat and Fred told me that Ted had hired another crewmember. I'm Jeff. Ted hired me too."

Jeff was a bit shorter than I am with preppie blonde hair. His speaking manner immediately projected a rare, non-antagonistic self-confidence. We shook hands when I introduced myself, careful to use only my first name. He smiled and said, "Since Ted hired you, I assume you went to college?"

I was a bit puzzled by the question, but answered, "Yeah, I graduated about a year ago."

His smile got even bigger. "Boy, am I glad you're here. Some of these guys on the crew really have a chip on their shoulder about anyone who got past high school. It's like they resent you because you tried to get an education. I don't look down on them or anything just because they didn't, but sometimes they just don't let up with their working class hero crap. At first I just went along with it, but now it's really starting to get old. It's nice to finally have someone else working onboard with aspirations past junior prom. Did you just get into town?"

"Yeah, just a couple of hours ago."

"Well, if you've got nothing else going on tonight I'll show you around, and maybe we'll get a pizza or something. Old town is kind of cool with all the strange little bars and conch houses."

I appreciated the offer; Jeff radiated a genuine friendliness. "Sounds great. Let me finish what I'm doing here and I'll be ready in a minute."

A short time later we were walking towards town across the *Galleon's* marl parking lot. Years later this property would be transformed into a high-rise luxury resort, but now just beyond the dusty parking area it resembled a seaside garbage dump. Jeff explained that this area had once been a boat yard but it had been abandoned and torn down with the debris simply left were it fell. As we approached the edge of the parking lot's street entrance, I noticed an old tan station wagon that had not been present earlier. A man was leaning against the front fender with a nondescript, overly fat dog at his feet.

Jeff called out to him as we approached. "Hey John, how's it goin' tonight?"

A coarse, loud voice volleyed back a string of garbled syllables indecipherable to my ear. Jeff seemed to understand the man's reply and said, "Yeah, it is hot tonight. By the way, this is Syd. He's a new employee so you'll be seeing him around here."

John and I shook hands. He was old and weather-beaten, the human equivalent of an abandoned tobacco shed. Now that I was closer to him I noticed a distinct lack of teeth when he smiled. He made a comment directed at me that I could not understand or interpret. Not knowing exactly how to respond, all I managed was a slight nod and lamely tried to look friendly. We briefly petted John's dog, said good-bye and walked down Front Street along the uneven sidewalk.

When we were out of earshot of the parking lot, I asked Jeff if he really understood what John had been saying to us. He laughed and said "Not everything, but enough to guess at the content. When I first got here I couldn't figure out what he was saying either, but after a couple of weeks I got better at it. I still avoid long conversations though."

"What's he do there anyway?"

Jeff's answer surprised me. "He's the security guard for the parking lot and the *Galleon*."

"Why would he need to guard the parking lot? The whole thing looks like a used nuclear test site."

Jeff shrugged. "I guess they had some problems sometime in the past with drunks climbing around on the *Galleon* at night. He's supposed to chase them off. Actually I think Mel might feel a little sorry for him 'cause he's an old retired shrimper who doesn't have much money. It's an easy job and hardly anyone comes around here at night. I think he leaves after one in the morning anyway."

We walked along Front Street to where it intersected with Duval Street and then followed Duval south. Both street and sidewalks were empty and the buildings reflected the sound of amplified music from indistinct sources somewhere further along. In 1979 Key West was not yet a major tourist destination and many of the shops along Duval were vacant or closed for the summer. Fifteen years later this same place would resemble a crowded carnival midway at night, but now it was just an ordinary small town main street. At Jeff's suggestion we entered a small pizza restaurant named Angelo's, ordered and sat down.

Jeff pointed across the street towards a building with "Sloppy Joe's Bar" emblazoned along the sides. "See that? That's the famous Sloppy Joe's bar where Ernest Hemingway supposedly used to drink. Except that the original Sloppy Joe's is actually around the corner and now called Capt. Tony's Saloon. From what I've heard, Hemingway never set foot in this place. You'd never know it from the way they advertise as "Hemingway's favorite bar". Sometimes though they've got good music; maybe later we'll check it out and see who's playing." When the pizza finally came, I asked Jeff how he had come to work for Treasure Salvors. Between mouthing overly hot bites of gooey cheese and sips of beer he recounted his story.

Sloppy Joe's, Duval Street

Jeff had been in ROTC in college, and after graduation had several months free before having to report to the Air Force for flight training school. He too had been captivated by the National Geographic documentary and had called the company about a job or volunteer work that he might do until having to leave for the military. When Ted hired him he had packed his dive gear and some clothes on the back of his small Honda motorcycle and rode it straight down to Key West from his home in Pennsylvania. The trip had been so uncomfortable that he strongly considered riding the bike off Mallory dock into the ocean so he would never have to see it again. Jeff had been here a couple of months and really enjoyed the job despite the low pay. There had been artifacts found on the last few trips to the wreck site, and he couldn't wait to go back out.

"You know, he said the reason that you and I and the others on board were hired is to replace the old crew who all quit. The guys who work on the boats now were hired right around the time that I was. Everybody's pretty new."

I reached for another piece of pizza and asked, "Why'd the old crew quit?"

"I guess most of them had been around at least a year or so and were fairly experienced. They were all getting pretty tired of the low pay and decided to form a kind of an informal union to force Mel into giving them raises. Well, Mel can barely make payroll as it is and told them that he would like to give them more but he just didn't have the money. They decided amongst themselves that Mel couldn't keep the operation going without them, that their experience would be irreplaceable. They could force him to give in if they all threatened to walk out. It turned into an ultimatum."

"Mel just went out and hired the first people he saw to replace them. That's how the crew we have got here. Most of them are just guys Mel met at a bar or happened to run into somewhere. But by hiring them, he kept things going at a level he could afford, and the old crew found out they weren't so irreplaceable after all."

I wiped my mouth with a crumpled paper napkin. "Well, I guess it's good for us that they quit, otherwise we wouldn't be here. By the way, how does Ted - I mean what's his job besides hiring you and me?"

Jeff shifted back in his seat. "Ted's the captain of the *Swordfish* and the operations manager. He decides which boat goes out to sea and what crew will be used. I think he and Mel discuss what's supposed to be accomplished before each trip and then Ted takes it from there. When I first came down here I thought that Mel went out to the site and directed things, at least that's what the National Geographic film implied. In reality Mel is too busy trying to get investors and keep the company surviving to go out to sea now. I don't think he's had the chance in years to spend much time on the site."

Jeff took a long drink from his glass and then reached for another slice of pizza before continuing. "Ted and Mel have a sort of an odd symbiotic relationship. They've got completely different personalities but somehow seem to work together. Ted is very authoritative and pragmatic and Mel is kind of an easygoing dreamer. Because of the two different personality types, two camps of alliance have formed amongst the employees as to who they relate to better."

Since I hadn't yet met Mel and my only real experience with Ted had been the exasperating job interview, I was curious who Jeff favored. "Who do you like working for best?"

Jeff paused before he spoke as if weighing his answer. "Mel's very easy to get along with 'cause he doesn't put many demands on the divers. Most of the guys like him since he doesn't hassle them much if they don't show up for work or come in late. He's always been nice to me and his enthusiasm is infectious; you can't help liking the guy."

"On the other hand, Ted's not quite so popular with everyone. He was a career officer in the Navy and retired with the rank of captain. After all those years of being in charge and ordering people around he still acts like he's in command. He expects everyone to be on time and to work all day like at a real job. I guess he wasn't real impressed with the work ethic or skill levels of some of the locals who Mel had hired, so he decided to start hiring people from elsewhere with college degrees. That's how you and I came to be here. He also has a favorable bias towards anyone who's been in the military. As you spend more time here you'll notice that a lot of the divers act like holdovers from the hippie days of the sixties. They don't like taking orders or real world job situations. That's where their rub with Ted is."

Jeff took a quick bite of pizza and then continued. "Don't get me wrong - everyone in the company respects Ted's knowledge, it's just that some of the guys hate working for him 'cause he doesn't cut them slack the way that Mel would. I personally like working for Ted; I guess I prefer his management style. He gets things done and as long as you do what he asks, you won't have any problems. He's kind of distant 'till you get to know him, but once you've made a few trips out to sea with him he kind of grows on you."

Jeff started squirming a bit in his chair as he finished telling me about Ted. With the pizza completely consumed, there was little attraction for either of us to stay any longer. I was about to suggest we leave when he said, "Say, we've been wasting time here talking about company stuff which you'll find out about soon enough anyway. Let's head down Duval and see if Bambi's dancing tonight at the Boat Bar. I met her a few weeks ago and I think I want to marry her and take her away from all this. She's just the kind of girl your parents would disown you over." We both laughed and walked out into the velvet night air.

Despite the late hour, when Jeff and I returned to the *Swordfish*, sleep did not come easily to me that night. The open hatch overhead and the tiny portholes offered some ventilation to the occupants of the upper bunks, but the lower tier where I slept was sweltering. Although the snoring that came from some of the crew didn't help, the most sleep-distracting sound came from the boat itself. There was an irregular set of odd noises that sounded like a metallic yawn, sometimes followed by an abrupt "tink". The metal hull of the boat just inches from my head carried the sound well but gave no hint as to its source. For a while I lay there and sweated, waiting for the dawn. Eventually I dragged the mattress outside and laid it on the starboard engine hatch. The cool night breeze carried me off to sleep.

At first light I woke up, covered in dew. Since nobody else on the boat yet stirred, I went over to the *Galleon* and followed the steps up to the highest deck on the sterncastle. From there I had a panoramic view of sunrise over the harbor. The waterfront's docks were burdened with what seemed hundreds of rust streaked shrimp boats. The density of their black, upward pointing booms and rigging gave the impression of a tract of burnt out forest. Several of them coughed out carbon laden diesel smoke as they clumsily maneuvered around two docks that projected from a complex of white concrete buildings. The buildings had the logo "Singleton Seafood" painted on them. In the morning air I could hear garbled voices over the company's outdoor P.A. system.

At the *Galleon's* stern and parallel with the stone jetty that protected the harbor was another curious fleet. These were old runabouts and small wooden cruisers that had been turned into a kind of floating shantytown. It would be obvious even to a casual observer that most of these vessels were in the final stages of dereliction. They were all moored independently of one another and had been heavily modified with scrap lumber and ratty tarps to form enlarged living areas on top of the listing hulls. At this early hour I saw no one yet stirred on any of them although there were indications that people lived on board.

The sound of gravel popping beneath tires announced two cars pulling into the *Galleon's* parking lot. When the old white Lincoln stopped, two young men in T-shirts and shorts carrying bags of groceries emerged along with the tall, skinny, red headed driver. They carefully trod up the dew-slicked gangplank onto the *Galleon*. The other car, a taxi, discharged two older men in long pants and sport shirts who followed carrying some lightweight soft luggage. They all stopped and talked briefly on the *Galleon's* weather deck and then climbed over the *Swordfish* to get aboard the *Virgalona*. One of the group made several more trips to the open trunk of the Lincoln and collected more groceries and dive gear. Before the last trip was completed, blue smoke and a garglely rumble at the stern of the *Virgalona* indicated its engines had been started.

I climbed down from my perch on the *Galleon's* sterncastle and crossed over to the *Swordfish*. With the *Virgalona's* exhaust note reflecting off the steel side of the *Swordfish*, it was hard for me to converse with those on board. The tall redhead was behind the wheel and pantomimed that he wanted me to release their lines from the *Swordfish*. That done, the boat slowly backed away, turned and headed for the harbor entrance.

Just as the *Virgalona* cleared the harbor and started a turn westerly, I was joined on deck by a yawning, scratching Jeff. He explained that the skinny red haired guy was Mel's son Kane who was the captain of the *Virgalona*. They were taking a psychic named Olaf and his manager out to the *Atocha* site to see if he could "sense" where the rest of the wreck was. The famous tabloid newspaper, for which Olaf worked, thought it would make a good story. We both chuckled at the thought and with the *Virgalona's* engine sound becoming fainter; Jeff stretched and went back to his bunk.

It was almost 9:30 before I saw anyone else on the *Galleon*. A stocky man with a white, full beard boarded the ship and climbed up the sterncastle steps from the main deck. In our brief exchange of greetings, he introduced himself as Bill and told me he was the company's bookkeeper. I followed him into his office which was in its own cabin in the upper sterncastle. His desk was centered between the large gallery windows at the stern and a thick column, which I assumed to be a lower portion of the mizzenmast. The office space would have made a convincing 1700's captain's cabin except for the stacks of water-stained cardboard boxes, filing cabinets and old bicycle parts piled

Bill the bookkeeper outside his office

about. We went through the mantra of filling out the appropriate tax forms and in short order I was officially on the company's books.

After I finished with Bill, I headed back down to the main deck and crossed over to the *Swordfish*. As I arrived on its back deck, Fred emerged out of the cabin entrance with a big smile on his face. "Hey man, I've been looking for you." I explained that I had been to see the bookkeeper and was now ready to go to work. "Good" he said. "I've got a job for you to do. You see, the *Galleon* is pretty old and the bottom is in really bad shape. I don't know if you noticed or not, but on the starboard side there are several large hoses sticking out of the hull with water pouring out of them. Those are bilge pump discharge hoses and they run nearly 24 hours a day. When the *Galleon* isn't leaking too badly, there may only be one or two hoses with water coming out of them. If it's really leaking, sometimes there may be up to seven pumps that will kick in. We've got to keep as few pumps running as possible. Mel doesn't want to take a chance on the *Galleon* sinking. Right now there are five pumps running, so you need to get into the water and patch up some of the holes."

"OK" I said. "How do you find the holes - I mean what do they look like and where are they likely to be?"

"They can be anywhere below the waterline, but they're mostly along the very bottom. You have to look closely as you move along the hull. Sometimes you can see stuff in the water being sucked in between the planks or sometimes an old patch gets knocked off when the *Galleon* hits the bottom at low tide. If there is a big hole you can actually hear the water rushing into it. We patch the holes with hydraulic cement, so I'll get someone to mix some up for you. All you gotta do is find a hole, clean the area around it with a scraper, and then come up and tell the guy who's mixing up the cement how much you need. As soon as he hands you the ball of cement, you swim back to the hole as fast as you can and push the ball into the hole and hold it there. It takes several minutes for the stuff to get hard, so you gotta hold it really still. Then you can go looking for the next hole. When you've fixed enough holes so that a couple of those pumps shut off, you can quit."

I looked over the side at the harbor water between the *Swordfish* and the *Galleon*. It looked pretty murky and the *Galleon's* large size meant that I would be examining a vast surface area with limited visibility. I wasn't overly confident in my ability to locate any leaks quickly. With that thought, I turned back to Fred and asked "Since I'm new at this, maybe it would be a good idea if you came in with me on the first dive. You could show me what to look for, and then I could find the holes quicker."

Fred's mild demeanor instantly changed. "I'm not getting in that nasty water! You're gonna have to pay your dues like everybody else! I got plenty of time under the damn *Galleon* and so do all the rest of us. You're the new guy, so it's your job now. No one held my hand when I was new and I'm not gonna hold yours. When they hire someone else, then it'll be his turn. Today starts yours, college boy. Get your gear on and get to it!"

Not wanting to antagonize Fred any further, I quickly got my gear out and got suited up. Joe, who was my appointed cement mixer, led me to a small platform next to the main gangplank between the *Galleon* and the shore. He had a bag of hydraulic cement, a garden hose and an old wheelbarrow in which to mix it. Using the discharge from the hose, he indicated a place on the harbor water's surface next to the platform. "If you jump in right there, you shouldn't have any problems. The water's about eight foot deep and there's no junk sticking up off the bottom."

I looked down at the water. Several dead rats floated in the thick mucous scum on the surface. When Joe had sprayed the hose on the water it had pushed some of the floating glop aside to form a relatively clear area into which I could jump. "What's all that crap in the water?" I asked.

"It's mostly discharge from the shrimp processing plant. It tends to collect in this part of the harbor. Nice, huh? When you come up, don't take the regulator out of your mouth until your head's well above the water."

There was no ladder that I could see. "How do I get out when I'm finished?"

Joe gestured with his arm. "Just swim over to the *Swordfish* and I'll put the dive ladder down for ya."

This certainly wasn't the type of diving I had visualized myself doing here, but as Fred had so emphatically put it, I was the new guy. It still beat the idea of working at my old office job. I put the regulator in my mouth, breathed several times and jumped from the platform into the receding area of clearer water.

When the flurry of bubbles from my entry had cleared, I swam slowly over towards the *Galleon's* hull. The water's visibility was about two feet and was surprisingly warm. Everything past my facemask was tinted by the tea coloring of the water. Though I had previously dove in the ocean, it had always been in bright, clear water near coral reefs and always with other divers. Now I felt isolated due to the poor visibility and lack of a diving buddy. There was a brief replay in my mind of every account I had ever heard of involving shark attacks or other dangerous marine animals.

I found the growth-encrusted side of the hull and examined it. There was nothing that I could see that indicated the potential source of a leak. I moved down and under the bottom in the direction of the keel. It got much darker as I went deeper. The *Galleon's* hull not only blocked out the light but the water itself was more turbid. To see the ship's bottom more closely, I turned upside down and followed the curve of the hull downward. Abruptly something snagged my tank and stopped my progress. With one hand against the *Galleon* I reached behind with the other and found that my tank had furrowed into the harbor bottom. I was stuck under it.

Before trying to back out, I relaxed for a minute to consider my position. As I lay there my eyes began to acclimate to the dim light and I saw the ship's keel about a foot away. Looking very closely I saw what looked like lumps of gray colored material stuck to the dark wooden planks next to it. If these were previous repairs, it would be easy to see how at low tide the ship's hull would hit the bottom and knock off some patches.

Suddenly I realized that if the tide were going out that this might not be the best place to linger. I had no idea what part of the tide's cycle we were in, but I pulled myself out from between the *Galleon* and the bottom as quickly as I could. During the course of the morning and early afternoon I went over the ship's whole bottom looking for obvious damage. There were some places where the wood had been badly eaten by shipworms, so I got some globs of cement from Joe and smeared it over those areas. The cement got hot as it cured and was uncomfortable to handle. Also, my regulator mouthpiece developed a leak, which allowed water into my mouth with every breath. The water didn't taste anything like normal seawater but I tried not to think about what I was ingesting.

Despite using up the air in several dive tanks and two bags of cement, my patching efforts didn't seem to have had much effect on shutting off any pumps. Wanting to finish the job, I carefully worked along the keel area again. Finally I found and patched an area near the bow where an older repair had fallen off. When I surfaced at the platform, Joe told me that he was out of cement and that at least one of the *Galleon's* bilge pumps had shut off. He indicated that we were finished for now so I swam towards the *Swordfish's* dive ladder. After I swam around the bow of the *Galleon*, I took the most direct route by going under the salvage boat rafted leeward of dockside. While I crossed beneath the *Swordfish*, a dense cloud of brownish particles descended over me. Later, Jeff would confirm the particle source: someone had flushed the *Swordfish's* head as I swam under the boat. A nice touch.

After I showered on the back deck of the *Swordfish* with a garden hose, I dried off and went below to put on a clean pair of shorts. With a hastily made peanut butter sandwich in hand, I reemerged on the back deck again and found Ted. He was examining an ancient looking outboard motor laying on the edge of the cabin roof.

"Where were you this morning?" He didn't look away from the motor as he addressed me.

"Fred told me to dive under the *Galleon* and fix some of the leaks. It took me awhile but I think I patched some. I just got out of the water a few minutes ago. Were you looking for me?"

Ted's gaze remained focused on the motor as he turned the propeller with one hand and worked its gear selector with the other. He suddenly spoke. "Postponing the inevitable, eh?" He paused again before continuing. "The gears in this thing don't work anymore. It's either jammed or there's something broken inside. One way or the other it needs to be taken apart. Why don't you work on this and find out what it needs. Tomorrow morning give me a list of any broken parts you've found and I'll pick up new ones for you."

I finished the afternoon sitting in the bright, hot sunlight, disassembling the motor. Despite the sweat that ran down my face as I performed this simple mechanical task, I was supremely happy. In Wisconsin I would have been sitting in my drab office cubicle, talking on the phone to people I would never meet while exchanging data on a computer terminal. There was never any sense of closure or resolution to the office work, and I had little feeling of accomplishment. It had been as if I worked on a vast assembly line installing a widget of unknown function on a device of uncertain usefulness.

Here the air was filled with the natural sounds of birds and the lapping of water. I saw the direct result of my labors and felt satisfaction in having fixed the problem. It occurred to me that this was the first time in my working life that I had felt a hint of job satisfaction… and in an exotic outdoor environment no less. Even better, we'd probably soon be out diving on the wreck site. My future felt as bright as the direct subtropical sun. Suddenly even patching the *Galleon* didn't seem so bad.

The next day was a carbon copy of the first. I dove under the *Galleon* in the morning and made repairs for Ted in the afternoon. During a break in the routine, Jeff offered to introduce me to Mel. We walked into the gift shop in the *Galleon* and through the doorway leading aft. There were several doorways on either side of the short, dim corridor that followed, but we continued straight back into the office entrance at the end.

At the back of the cramped, wooden walled office sat two desks sitting side by side and facing the doorway. They were flanked on all sides by filing cabinets and overly burdened shelving. The air precipitated mildew and cigarette smoke despite the best efforts of a small air conditioner that wheezed feebly in the wall. Both people sitting behind the desks were instantly recognizable to me from the National Geographic documentary. Mel Fisher read some documents at the desk on my left. His wife Delores smiled as she looked up at us from her desk on the right.

Jeff stepped forward slightly. "Hi Mel and Deo. I want to introduce you to Syd… he was hired a couple of days ago by Ted. He's been pretty busy around here since he started and hasn't had the chance to meet you guys yet."

I remembered, as she said hello to me, that Delores was called "Deo".

I returned the greeting. "Hi, nice to meet you. Your voice sounds familiar, maybe it was you I spoke to several weeks ago when I was inquiring about a job here." She had nice reddish hair and a very pleasant smile. Though middle aged, her facial features still implied the good looks of her youth. I had been impressed reading about her diving background, from setting underwater endurance records to recovery diving on the various wrecks on which she and Mel had worked.

Mel took a squinty eyed drag from his cigarette as he leaned across his desk to shake my hand. "How's it goin." He spoke in a slow, nasally fashion. Even though he sat, it was obvious to me that he was a very tall man. The open collar of his polyester shirt framed a gold coin mounted on a chain around his neck. Though not fat, his stomach pressed hard against the lower fabric of his shirt.

I tried my best to be charming. "It's great to meet you both in person. Like a lot of other people, I've followed your company's story for some time. I'm really excited about working here and hopefully I can help find the rest of the *Atocha* with you." The four of us talked until the conversation ran thin and then Jeff and I said our goodbyes and exited the office. On the way back down the dim corridor towards the gift shop, Jeff stopped at another office entrance on the left. "There's one more person here you should meet before we go." He opened the door and I looked in to see a woman behind a desk in another munchkin office space.

Jeff spoke first. "Hi Leah, I wanted you to meet the latest escapee from the Midwest. This is Syd Jones and your husband hired him, so if he doesn't work out you know who to blame."

Leah stood and shook my hand with a big smile." Nice to meet you Syd. I don't know if I should welcome you or warn you to get away while there's still time. I came down here to help Deo in the office once, and next thing I knew I was working here forty hours a week." She had a friendly, effervescent personality and speaking style. Leah appeared to be in her late forties, had blonde hair and wore glasses She was much shorter than I was.

"Is Ted really your husband?" I asked.

"That's what he keeps telling everybody" she grinned. "Remember though, he may be a captain but I'm the admiral!"

Tropical Depression

After several days, the *Virgalona* returned from her trip. Using small styrofoam buoys, the boat's crew had marked a number of areas that Olaf's "psychic powers" told him contained gold from the *Atocha*. There was plenty of smirking amongst the *Swordfish* boys about *Virgalona's* mission, until she went back to the site again the next day, leaving us again with the responsibility of keeping the *Galleon* afloat. I did however receive my first paycheck - it turned out to be $30.00 instead of the $50.00 I had been promised. The rest of the employees had also suffered the same shortage. Apparently there just wasn't enough money in the company to go around. The checks were doled out by Bill the bookkeeper with the promise of "catching up" on our missing pay when more money came in. By the way the rest of the crew accepted the news, I assumed this was a fairly frequent occurrence.

There was a flush of excitement amongst the senior crewmembers of the *Swordfish*. A local businessman had made a deal with Mel to use his own vessel in an attempt to locate the *Atocha*. The vessel was a large seagoing tug named the *Jefferson* and it had just been refitted in the boatyard. The crew's quarters reportedly were paneled like a fine yacht and featured hot water, air conditioning and even a washing machine. Part of the deal stipulated that Treasure Salvors would supply the necessary experienced dive team that would live and work on board.

Fred was especially self-impressed that he and some of the others had been chosen to work aboard the *Jefferson*. Though the *Jefferson* had at least two weeks of preparations before she picked up the dive team and headed for the site, Fred's current interest in his work duties aboard the *Swordfish* grew less the more he boasted about the *Jefferson's* attributes. He reminded us that compared to the luxurious accommodations aboard the tug, the *Swordfish* and the *Virgalona* were little better than the dirtbag derelicts anchored off the rock jetty in the harbor. The *Jefferson's* powerful engine and gigantic propeller could force huge volumes of water through its specially designed prop wash deflector. The deep sand that covered the *Atocha's* remains would easily be pushed aside by the thrust of so much horsepower. It was just too bad that those of us left aboard the inferior company boats would miss out on the big finds sure to come.

I found Fred's self-assured cockiness about the *Jefferson's* chance of success very deflating. For the last week, we had done little else but more underwater patching of the ever leaking *Galleon* and the occasional mechanical repair job on the *Swordfish*. During this time Ted had made no mention about the *Swordfish* making a trip to the *Atocha* site, and I noticed that the rest of the crew seemed leery of asking him about it. He would show up every day about the same time and disappear into the *Galleon's* office area for a while. When he reappeared he stopped at the *Swordfish*, gave a few terse orders to the assembled crew and then left just as abruptly as he had arrived.

Ted's intimidating nature was especially effective when dismissing troublesome employees. One morning he briskly walked just past his intended target, abruptly stopped and with a brief turn growled deeply, "Get your shit off the boat!" There was no question of negotiation or second chances. Ted's body posture was definitive enough.

Once I got up enough courage to ask him when we might be going to the *Atocha* site. He looked surprised, put his index finger in his mouth to wet it and held it in the air. "Feel that?" He indicated the wind. "When it stops blowing, we'll go out."

My impatience carried the conversation further than his vocal tone deemed prudent. "It doesn't seem too windy to me. The waves outside the jetty look pretty small, and the *Virgalona's* been out several times."

Ted stiffened noticeably at my challenge to his answer but then smiled slightly. "Forty miles out to sea it'll look a lot different. When the weather's good, we'll go... and not before. The *Virgalona* burns a lot less fuel than we do and we don't have the money right now for both boats to go out. If it makes you feel any better, we're taking the *Swordfish* over to the fuel dock tomorrow. We'll be taking on diesel so that we're ready when the wind does die down."

Later that afternoon I got my first tour of the *Jefferson*. When our work for the day had been completed, Jeff suggested we stop by and visit Fred. Lately he seemed to spend most of his time involved with unspecified but important duties aboard the tug. We had both noticed that Fred had been involved with work activities around the *Swordfish* only when Ted was present, and left quietly after he departed. If asked by another crewmember where he was going, he replied only that his expertise was needed for some project aboard the tug. Jeff had been aboard the *Jefferson* a few days earlier and offered to take me over to it. We both were a little curious as to what Fred did over there.

The *Jefferson* was tied up at an old navy pier a few blocks away from Treasure Salvors. As we approached her, the fresh paint, which shimmered in the wave, reflected sunlight and a neat overall appearance complimented its classic seagoing tugboat profile.

The tug *Jefferson* with its huge propwash deflector

The boat had been modified with a large "A" frame mounted above the stern that supported one end of a mammoth rectangular aluminum propwash deflector. The other end of the deflector was hinged to the stern of the tug at the water line just above the rudder. The deflector was similar in concept to the "mailboxes" of the *Swordfish* and *Virgalona*. The end supported by the "A" frame via rope and pulleys could be lowered into the water and secured pointing downward towards the sea floor at about a forty-five degree angle. With the vessel's position held by a web of anchors, the main engine's propeller thrust would be forced downward by the deflector, pushing away the loose sand covering the wreck site. The cumulative visual effect of the pristine vessel with its purposefully built excavation equipment seemed only to confirm Fred's confident boasting.

There was no sign of the crew and the only noise was the flat monotone exhaust discharge of the tug's generator. We rappelled down a homemade wooden ladder from the concrete pier to the *Jefferson's* main deck. Located midway along the deckhouse was a watertight door. As Jeff opened it, a cascade of cold air carrying a loud rattling noise and music engulfed us.

We stepped into an expansive galley defined by polished stainless steel, white Formica and varnished wood. The air conditioning was so cold we could almost see our breath. Next to a counter across the room, Fred and a large, bearded fellow intently watched the contents of a spinning blender. Immediately surrounding the blender were several liquor bottles and pieces of fruit. Fred glanced in our direction, smiled and nodded before he turned off the blender. It was then I realized how loud Jimmy Buffet was cranked on the stereo. "Hey Jeff, Syd. We're mixin' up some tropical drinks here. Nothin' nicer onna hot day. You want some?"

Jeff spoke up. "Sure, OK. We just stopped by so I could show Syd the *Jefferson*."

Fred introduced the bearded guy. "This is John, he's the first mate on board. He's a good man and we're gonna find a ton of gold together on this boat!" It was immediately apparent to me from in the looseness of Fred's speech that the blender had been working most of the afternoon.

Fred left John to mix up another batch of drinks as he led me on a tour of the various compartments. As we moved through different areas, he described the tug's features in a way that implied a pride of ownership. Even with Fred's obvious bias towards the *Jefferson*, it was hard to not be impressed. All the inner spaces, including the engine room gleamed and looked very yacht-like.

Later that evening as we walked back towards the company boats, Jeff and I talked about how different life was on the *Swordfish* compared to life the *Jefferson*. On the *Swordfish* we took showers and brushed our teeth with a garden hose on the back deck, often in full view of the patrons of the nearby A&B Lobster House restaurant. Though there was a shower stall in the crew quarters down below, it had no working water supply or drain. There was no hot water for cooking or bathing, and sleeping at night was often difficult in the hot, uncirculated air of the mildew farm/bunk area in the bow. Piles of dirty laundry with the crew's personal effects were casually scattered about. Since batteries instead of generators supplied electrical power, the few interior lights were dim and could only be used for short periods of time. If the *Jefferson* was like staying in a luxury hotel, the *Swordfish* was similar to an extended camping trip. The neglected *Virgalona* was even worse.

The next morning Ted arrived earlier than usual. He said nothing as he opened one of the large engine room hatches on the *Swordfish's* deck, descended between the engines and began checking the fluid levels. I squatted next to the opening and watched with a look of reverent silence. After he finished the inspection, he climbed out, shut the hatch, and approached the stern control station at the aft port side of the deckhouse. With his index finger he pushed one of two black rubber buttons mounted next to the steering wheel.

Twelve pistons instantly slammed in their cylinders, punching a drum roll of cannon shot decibels and blue smoke out of a ten-inch exhaust pipe at the stern. The exhaust noise was so loud we could hear it through our lungs. Ted pushed the other black button and twelve more pistons in the other engine challenged the shout of the first. Up until this moment, the *Swordfish* had always seemed an inert object to me. Now it coughed, breathed and shook out stiff muscles.

Before the hour was up we had left the side of the *Galleon*, gone to the fuel dock across the harbor and taken on 500 gallons of diesel fuel. On the short trip back to the *Galleon*, I sat on the forepeak of the bow and felt the steel under me vibrating with the working of the engines. At last it looked like we were making preparations to go to the site. Just as the *Swordfish's* engine noise overpowered normal conversation, this short trip dispatched my growing distaste for life patching the *Galleon* and listening to Fred. Our trip to the fuel dock suggested a near future change in the routine. Maybe soon I would get my chance to see the site.

A late summer tropical wave moved through the Keys, causing high winds and almost continuous rain for several days. The *Virgalona* came in from sea, and her crew, like most of the *Swordfish's*, spent their idle time drinking in the various nearby bars.

I was still doing most of the daily *Galleon* repair diving, made even less appealing by the knowledge that the rest of my shipmates were off having fun. The nonstop alcohol consumption amongst the crew had other unpleasant effects. Arguments seemed to break out easily between the participants. When they periodically returned to the *Swordfish* to raid the food pantry, they prepared and abandoned meals without any effort to clean up. I sometimes returned to my bunk area after diving to find greasy pots, pans and dirty utensils, which overflowed from the galley sink onto my bunk.

It was irritating, but I rationalized away this behavior. Though fragile, there was an implied camaraderie amongst the crew - and being the new guy I was still trying to find my place amongst them. Activities and behaviors that I would have normally avoided or taken exception to were excused for now in the interest of getting along with my new colleagues. This desire for acceptance had nearly fatal consequences.

One afternoon during a break from *Galleon* diving, Jeff and I talked about seeing a movie playing at a theater in new town later that night. Fred, on a now rare visit back to Treasure Salvors from the *Jefferson*, overheard us talking and asked to come along. As I had the only car between us, it looked like I would be responsible for transportation. Since the movie didn't start until ten o'clock, Fred asked that we pick him up at the Boat Bar when it was time to go. He had plans to play pool there during the early evening.

At nine thirty that night Jeff and I stopped as planned to pick up Fred. The Boat Bar was a poxy, loud music joint at the corner of Duval and Eaton streets. People drawn to it were the type who avoided fern bars the way a vampire avoids daylight. Its breath of stale beer, vomit and urine spewed out onto Duval every time someone opened the door. In the front room were several worn pool tables across from the bar. In the back, topless dancers enticed transient redneck shrimpers to spend their money on tips.

As we entered, it somehow seemed darker inside than the night was outside. We both strained to spot Fred's face amongst the crowd at the pool tables. Jeff and I were both wearing white T-shirts with "Atocha Expedition Crew" emblazoned on them. Their clean, washed appearance contrasted strongly to the apparel of most of the patrons.

A group of six individuals drew out of the crowd and formed a circle around us in the din. Their leader, slightly shorter than me and with a very aggressive posture pushed close to us. "You guys Fred's friends?"

Jeff answered first. "Yeah, we're here to pick him up - where is..."

Before Jeff could finish, the leader pushed slightly closer. "Fred didn't play pool too good tonight. He owes me a hunnerd and fifty bucks. He said he didn't have the money, and that you guys would pay for him. I want my hunnerd and fifty NOW."

I sensed somebody standing right behind me. Jeff started a reply but I talked over him. "Hey, we're just picking Fred up to go to a movie. We don't have anything to do with this. We don't have any money - whatever business you have with him is between you two... not us. We just work with the guy."

The circle got smaller as the leader's muscles tensed. "He said YOU were gonna give me the money - I want it NOW!"

Jeff answered this time. "Listen, he probably went back to the boat. We'll go find him and tell him that he needs to square things with you guys. Why don't you come down there in about an hour and we'll make sure he's there so this all gets worked out."

The leader paused for a second and relaxed just slightly. "I know where your boats are. We'll be there in exactly one hour. Fred better be there - with the money and no bullshit or we're gonna fix him good."

So much for the movie. It only took us about ten minutes to get back to the *Galleon*. When we went aboard we found Fred loitering at the main deck railing, quite drunk. We confronted him about our encounter at the Boat Bar. Without apologizing for involving us, he started raving in drunken logic. According to his fuzzy sounding interpretation of the rules, on the last shot his opponent had done something that negated Fred's entire debt. Besides that, he had spent all his money drinking.

We told him that his opponent and his friends were going to be here shortly and he better figure out how to deal with the problem. With that, Fred went aboard the *Swordfish* while Jeff and I remained on the *Galleon's* deck. I assumed he was digging through his possessions, looking for more money.

When Fred returned, we were surprised to see that he carried a large brass flare pistol. This was not the kind of gun that is commonly carried by recreational vessels, but a large caliber, powerful old military style gun. He loaded it as he approached us. "If those bastards show up here I'll give 'em somethin' alright." He started cackling after a long swig from a beer bottle.

The situation had clearly escalated. From the dark of the *Galleon's* deck we could see several blocks down Front Street. Sure to their word, the group of six, silhouetted by the light from the dim yellow streetlights, walked in our direction. Several of them appeared to carry lengths of pipe or boards.

Fred was unmoved as the threat approached, and remained drunkenly defiant. Jeff and I tried to reason with him, improvising as we went. As the six entered the far end of our parking lot, we finally convinced him to sort out his problems with the leader. Jeff and I would talk to them first, hoping that we could convince them that only the two directly involved in the pool game should be involved. All the rest of us should stand down off to the side. Fred agreed to leave the flare gun on the *Galleon*.

As the group approached the *Galleon's* gangplank they appeared even more threatening. Since they obviously carried weapons, Jeff and I looked for something to take with us before we went to meet them. I found a broken spear gun lying next to the *Galleon's* railing; hopefully in the dark they wouldn't notice that it was non-functional. With a quick nervous glance at each other, Jeff and I walked down the gangplank toward them while Fred stayed out of sight.

As they saw us approach, they stopped and when we got close enough they started to form a circle around us as in the Boat Bar. We talked with forced calmness to the leader. Though nobody put down their weapons, he agreed that those of us who were not directly involved should walk away and let him and Fred resolve their differences.

With the agreement, some of the tension eased. Jeff called for Fred to come out and meet the leader in the parking lot. From the dark *Galleon*, Fred appeared on the gangplank empty handed and strode confidently toward us.

As he came closer, the leader motioned for him to follow him in the direction of the well-lit A&B Lobster House parking lot across the street. The rest of us would stay where we were. As Fred passed our group, one of the leader's friends suddenly jumped at Fred's back and shouted a warning.

"He's got a gun!" The leader's friend grabbed the handle of the flare gun, which protruded above the belt in the back of Fred's pants. As the gun came out, Fred turned and sprinted back in the direction of the *Galleon*.

The leader grabbed the gun from his buddy and pointed it at Jeff, who was about five feet away, backing slowly in the direction of the *Galleon*. He pulled the trigger and shouted at the same time but the hammer did not move. "God DAMN!" With that, Jeff started to turn and run. Quick as thought, the leader pulled the hammer back and noticed that I had frozen in my tracks next to him and still held the broken spear gun. Before I could drop the useless weapon, he had pivoted and pointed the short barrel right in my face. I saw the flare cartridge down in the muzzle, illuminated by the streetlights behind me.

He pulled the trigger. The hammer fell with a click. Nothing happened.

He cursed as I started running with Jeff towards the *Galleon*. The rest of them followed and screamed curses at us. Suddenly Fred reappeared in the parking lot. He carried his guitar and shouted "I'll pay up, I'LL PAY UP!"

Next morning, I sat alone on the back deck of the *Swordfish* and sweated in the breath of the tropical sun. As the harbor around me awoke, I relived the previous night's fiasco.

Fred had at the last moment defused the situation when he offered his guitar and $75.00 to his pool opponent. Had he not done that, it was hard to imagine how much worse things might have gotten. With the acceptance of Fred's offer, our adversary and his buddies were all handshakes and back slaps with us as only real drunks can do. They had even willingly given back the flare gun.

After they had left, Fred showed us that the adjustable firing pin on the gun's hammer had been backed off, which accounted for why the flare had not gone off. He then tossed the pistol into the harbor and we had all gone to bed.

It was time now for me to reevaluate my job at Treasure Salvors. After last night, any illusions I had about pursuing the acceptance of my "shipmates" or a perceived "esprit de corps" were over. In reality, Fred's little episode was only the latest in a never ending series of inane activities on which many of the crew seemed to thrive. Several were in a perpetual cycle of drinking/drugs, arguing, fighting and then disappearing for days. Most had little real thought beyond their next paycheck and the odd possibility of going treasure hunting in between drunken antagonisms.

Oddly, short-term visitors or tourists were enthralled with these types, and saw them as "colorful" or "local characters". They were really nothing more than just non-skilled drifters, dirtbags, or rough ex-shrimpers who worked occasionally for the fifty bucks a week that Mel could afford to pay them. It would be interesting to see just how "colorful" the tourists would find these guys if they had to live with them 24 hours a day.

When the experienced previous crew quit over a money dispute, the sudden personnel vacuum had to be filled quickly to keep the company going. The suction had not only drawn in good guys like Jeff, but also every unemployed loser within stumbling distance of the *Galleon*. And sadly for both the company, and me, Jeff would be leaving soon for Air Force flight training.

I decided that if I was going to stay, that I would spend a lot less time trying to interact with the other divers during off times. From now on I would just focus on getting what I could out of the job. Maybe things would be different if only we would just go out to sea. I was determined to at least make one trip to the site before deciding whether to move on or not.

Miguel's Navy

The *Jefferson* had made a brief trip to sea to test the operation of its new propwash deflector. The deflector had worked well, but while being winched up at the conclusion of the test, the "A" frame that supported the structure collapsed from the strain. After several days of welding repairs, the *Jefferson* was at last ready to depart on its first official working trip to the *Atocha* site. The best news for me was that Ted was going to take out the *Swordfish* and her crew to help position the *Jefferson* in the best spot to use its powerful excavation abilities. Then we would electronically survey the surrounding area to try to find more of the *Atocha's* wreckage.

Since our trip would last at least four days, food had to be bought, and propane tanks for cooking had to be filled. Engine and transmission oils, water levels for coolant and batteries were all checked. Since the *Swordfish* had no compressor for filling empty dive tanks on board, Ted removed the twenty empty tanks from the tiny onboard dive locker and took them to a local dive shop to be filled. My orders were to double-check the working condition of the near-fossilized outboard motor on the old Boston Whaler we planned to tow behind us.

While Ted had gone to fill the dive tanks, Wally the cook arrived from the store via taxi with a giant load of groceries that took four of us several trips each to unload. He bought plenty of snacks in the form of cookies and pound cakes that were already being consumed during the transfer from taxi to the boat. The food that wasn't eaten immediately was stowed down below in the galley and in the large fiberglass walk-in cooler under one of the main cabin bunks.

Jeff, who had been drafted to help crew the *Jefferson*, had already left with his dive gear and belongings to join her. With Treasure Salvors' diver roster at a minimum, there weren't enough people to send out the *Virgalona* as well. Even if there had been enough spare crew available, the company still just didn't have enough money to support operating the two vessels at the same time.

When Ted returned with the dive tanks, we unloaded them from his pickup and put them back onboard in the dive locker. After he brought aboard a small duffle bag with his own possessions, Ted briefly questioned each of us about the status of the rest of the preparations. Apparently satisfied with our answers, he silently went to the port wheelhouse door, unlocked it, and went in.

Ted outside the wheelhouse door of the *Swordfish*

As this was the first time I would spend more than a just a few minutes in Ted's presence, I watched the rest of the crew closely for cues on how to behave around him. He wasn't exactly unfriendly, but his bluff mannerisms suggested someone who liked to both start and end any conversations. Even dressed in his old plaid shirt and faded pants, the timbre of his voice and short, measured speaking style imparted an authoritative stature. Years of navy officer experience had certainly left their mark.

I noticed that the rest of the crew had pretty much stopped all movement and conversation while they quietly looked at the wheelhouse. Ted had opened the starboard wheelhouse door from the inside as well as the adjacent sliding windows. He stood behind the wheel, and for an instant I could hear the geared warble of a starter motor briefly before the concussion of unmuffled exhaust gasses punched the atmosphere. Seconds later, the second main engine doubled the bass gargle. As the boat idled, the air around the stern blurred with steam and diesel smoke.

Even with the noise, none of the crew had moved and still stared toward the wheelhouse. After about a minute, Ted leaned partway out the wheelhouse door and pointed to the spring lines (ropes that hold a boats fore and aft position while docked). The nearest crewmember quickly untied them. Ted then pointed to the stern and then the bow; those lines were quickly released in that sequence. On October 19th we pulled effortlessly away from the *Virgalona* and the *Galleon*, and slid over to the docks at Singleton Seafood.

Since I assumed we were heading directly out to sea, I was surprised to see that we tied up again at the shrimp dock. Once the lines had been secured to the dock and the engines turned off, Ted directed me to help the others get the ice hose onboard. The ice hose was a green corrugated semi-flexible plastic pipe about ten inches in diameter and about forty feet long. One end was attached to a rigid metal pipe that lead from the nearby icehouse. We wrestled the stiff plastic hose through the main cabin door and into the open doorway of the walk-in ice cooler. At Ted's signal, the dock attendant shouted an unintelligible command through an intercom on the dock. Wally, standing next to me as we held the pipe in the doorway said "Hold on tight. This thing really kicks."

The sound of a loud throbbing buzz saw vibrated through the pipe. After a few seconds the whole tube bucked and squirmed as hundreds of pounds of crushed ice was blown through by a stream of compressed air. The cooler, which measured about eight feet long by five feet wide and high, was nearly full by the time the ice flow stopped two minutes later. We then struggled with the now even stiffer hose as it was removed from the boat and laid back out on the dock. The icehouse normally supplied fishing boats with the ice to keep their catches from spoiling. Since Treasure Salvors' boats at this time had no working generators and no electric refrigeration, the stop at the ice dock was a regular routine at the beginning of every trip to sea. Food that needed to be kept cool was simply thrown onto the ice just inside the cooler's door.

Minutes later we were again underway and left the harbor. As we passed into the channel between Key West and two small spoil islands just to the north, Ted advanced the throttle foreword and the *Swordfish* seemed pushed ahead as much by decibels as by propellers. The increased noise from the stern exhaust outlets was so extreme that everyone onboard automatically moved as close to the bow as possible. Two of the divers settled up on the foredeck ahead of the wheelhouse, sitting in coils of line there as if they were beanbag chairs. I followed Wally and a diver nicknamed "Mr. Science" into the confines of the wheelhouse and occupied the two large padded bench seats to the rear and on either side of the steering station.

Ted glanced in our direction as we entered and said "Wally - take the wheel and steer two four zero". Wally, his ego stroked at having been chosen over the rest of us, replied in his best gravelly pirate voice "Two four zero! Aye, Cap'n."

While Wally steered, Ted pulled out a hardbound book from behind some of the electronic equipment mounted on the console. The cover was dark blue and had the outline of a gold swordfish engraved on it. I watched him write several entries in it and then skip to the last pages where he began to work out a geometry problem. He seemed totally immersed in his efforts and ignored the conversation between Wally, Mr. Science, and me. Every now and then, without looking up, he ordered Wally to change to a new compass steering course.

Once Ted finished with his calculations, he stood up and began to enter numbers via a keypad into a device mounted on the console. It was a plotter, a machine that could indicate the course that the boat was taking on a built in nautical chart. The plotter was wired to a LORAN-C receiver used for electronic navigation. Using position information from the LORAN, the plotter drew the *Swordfish's* progress on the chart with an electronic pen.

As the wheelhouse was normally locked in port and was off limits to the crew, this was the first opportunity I had to really look around and inspect the equipment. The console and engine instrument binnacle were all made of mahogany that was shedding blistered varnish. There were two rather substantial looking metal fathometers (electronic water depth sensing equipment), another smaller LORAN-C, and a VHF radio for communications. On a padded shelf behind the bench seats sat several pieces of large electronic equipment I did not recognize. The steering wheel Wally held was made of chromed tubing with small wooden handles that projected outward from the rim.

As we made our way westward, we dodged the small colored lobster pot buoys that tensed at the ocean's shrug. About two miles north and parallel to our course were the continuation of the low, mangrove dressed Keys. Many visitors to Key West assumed that it is the last of the island chain, however the Keys continued on in irregular spurts for another ninety miles, terminating at the Dry Tortugas. The largest area of open ocean between the lower Keys is between a circular island group called the Marquesas Keys and the Dry Tortugas. In that gap was a shallow area of underwater sand dunes called the "Quicksands." The **Quicksands** was where Treasure Salvors had found the first artifacts from *Atocha*. The *Swordfish* took nearly three and a half hours to get to the site from Key West.

After about an hour at the wheel, Wally requested relief so he could start work on lunch. I quickly volunteered my services as a replacement. Ted granted my request, gave me a course heading and I took over for Wally.

The *Swordfish* steered easier than I had expected, in fact I had to be careful not to overcompensate if I got off course. To keep Ted from noticing my wandering path through the water, I began to question him about the various islands in view to the north of us. He seemed to know quite a bit about them, and had a separate story about each one.

As we passed south of the Marquesas Keys, he told me about how the Spanish salvors had established a camp there not long after the *Atocha* and the *Margarita* sank. The camp had been their base of operations for several years while they tried to locate and recover the ships. The modern name, Marquesas Keys, was a derivation of the name the original Spanish salvors had bestowed upon them at that time; Cayos del Marquis. The islands still looked much as they did then, populated only by birds, biting flies and mosquitoes.

Past the Marquesas was an unbroken horizon of open ocean. The water's color began to change from the greenish tint near the islands to a deeper blue. Another hour's travel and we were completely out of sight of any land.

Even from a distance, the *Arbutus* looked like a shipwreck.

Directly ahead of us, a mirage-like form struggled to gain substance. As we approached, the dark object morphed into the shape of a large vessel. Ted took the helm from me, pulled back the throttles and instructed us to get the mooring lines ready, as we would be tying up shortly. At first sighting I assumed that we approached the *Jefferson*. Now that we were less than a mile away, the profile left no doubt that we were visiting a ship about which the other divers had told so many horror stories. It was the *Arbutus*.

The *Arbutus* was a 187-foot long World War II era ex-Coast Guard buoy tender. The government had decommissioned it, and in 1975 Mel had bought the stripped out hulk in Miami and towed it to the site to use as a workstation and living accommodations for the divers on the *Atocha* project. There originally had been ambitious plans to try and refit and repower the beast, but the financial realities of Treasure Salvors had eventually nixed the idea. After being used for a time as a platform for an excavating device called a "Hydraflow", the ship quickly deteriorated into a floating wreck.

In port, one of the first things which gloating senior divers told new hires about was the *Arbutus*. The ship was now used only to keep a physical presence on the *Atocha* site. It was always kept manned by a single individual whose job it was to keep an eye out for any non-company boats that might try to dive or otherwise conduct salvage work in the wreck area.

Since *Treasure Salvor's* boats spent much of their time in port due to bad weather or lack of funds, the diver who was stationed on board the *Arbutus* was usually stuck on it alone for weeks at a time. Senior crew members loved to point out that like patching the *Galleon*, the newest employees were always picked first for *Arbutus* duty.

Jeff had told me stories about the poor condition of the ship, but as we drew alongside, it turned out to be much worse than I had imagined. We tossed our lines to a really dirty and rather obese crewmember on board, and as soon as we were secured, Ted cut the engines.

"Hey Maurice, you still here?" Wally cackled as he handed bags of groceries over the *Swordfish's* railing to the *Arbutus'* occupant. "One day when you've paid your debt to society, we'll let you come back to Key West. First, you're gonna hafta get cleaned up though. Man, I don't know how you can stand it out here for so long."

Maurice was a special case. Even though he wasn't a diver, he had convinced someone in the company to hire him as the *Arbutus'* caretaker. Since he was only paid when he was on board the ship, his employment with Treasure Salvors had been one of Robinson Caruso isolation. For months he had mostly been at sea by himself. The loneliness and boredom only seemed to bother him once in the while. When this happened he was taken back to Key West for a few days as a break. One of the new divers was left on board to take his place until he (hopefully) decided to come back.

Some of the crewmembers on the *Swordfish* joined Wally and started tossing bags of groceries and supplies over to the *Arbutus* faster than Maurice could take them. I took that as my cue and jumped aboard to give Maurice a hand. My real motive though was to get a closer look at the ship

As I climbed over the rust blistered gunwale, the first thing I noticed was that every surface I touched left reddish, greasy stains on my hands and clothes. As I grabbed the grocery bags that came over from the *Swordfish*, I also got a better look at Maurice. His clothes, hair and skin were thoroughly tinted with the same rusty smears I now had.

Alongside *Arbutus*

When the grocery transfer was finished, I asked Ted for permission to explore the *Arbutus* a bit further. As he adjusted the position of an old truck tire tied to the side of the *Swordfish* for protection against the bigger ship's flank he replied, "You've got ten minutes. Any longer and you're on board for the duration."

Along the passageway there were openings in the deckhouse where portholes and doors had once been. The doors themselves and brass framed glass portholes were now long gone, and the remaining apertures had the warmth of empty eye sockets in a skull. Fragments of red and white paint were still visible in places higher up along the deckhouse, but the reality of the *Arbutus* was rust. This was not a casual surface corrosion but a deep and complete festival of iron oxide. Huge craters flowered layered petals of decay with still more boils of metal decomposition pushing upward on every surface. As an added garnish, everything was slightly slippery from a coating of salt spray and grease.

I followed Maurice with armfuls of grocery bags into a doorway midship and stepped into what was once the galley. There were some rusted stainless steel counters and sinks still bolted to the deck but the rest of the original cooking equipment was nowhere to be seen. A large battered Igloo cooler with some ice in it served as temporary refrigeration for perishables, the rest of the food stocks joined the rusted, lableless cans stacked on some homemade shelving along the foreward bulkhead. While Maurice tended to his housekeeping duties, I continued my walking tour.

Aft of the galley was another doorway that lead to the engine room. With no machinery left there, it was just dark and cave-like, a deep grotto with rusty pipes and catwalks like vines leading into blackness. Nothing there invited further investigation.

I climbed up a ladder and wandered through vacant cabins just aft of the wheelhouse. The decking in these compartments was in very poor shape and in some places large holes had completely rusted through the floor. Another short flight of steps and I found myself in the wheelhouse.

The wheelhouse had a panoramic view of the front deck and bow of the *Arbutus* but no longer contained the ship's wheel, instruments or anything else. Discolorations on the flooring and bulkheads still hinted where equipment had long ago been mounted. Now the space was as open as a dance hall except for what had once been a wooden chart table positioned near the starboard door. Laid out on the chart table was a crumbling, naked foam mattress layered under a pile of filthy clothes, sheets and coverless paperbacks. I had located Maurice's nest.

The *Swordfish's* engines suddenly started. As the noise echoed through the empty metal canyons of *Arbutus*, I ran down to the main deck. Maurice was already being given the signal by Ted to throw off the lines as I jumped aboard. I helped the other divers pull in the ropes as they were released and coiled them neatly on deck. As we pulled away, one of the crew shouted to me above the noise; "Ted talked to the *Jefferson* on the radio while you were on the *Arbutus*. They had some sort of problem right before they left Key West and got a real late start. It looks like we're going to meet them out on the site tomorrow. Since the day's pretty well wasted, we're heading for the Marquesas to anchor for the night."

It took us an hour and a half to get to the Marquesas Keys from the

Arbutus. As we neared the islands, I stood half in the starboard wheel doorway and watched Ted as he guided the boat through the shallowing waters. Watching the LORAN-C and the fathometer like a pilot flying on instruments, he idled the *Swordfish* towards the beach near the southwestern most island. As I looked over the side, I could see the sandy bottom moving under us. It looked so shallow that I braced myself in the doorway and expected an impact, which never came.

There was a lone palm tree on the beach that we approached. About five hundred feet off the beach and abreast of the palm, Ted stopped the *Swordfish's* headway and shouted to the two divers who stood on the bow to drop the anchor. He then turned to me and said "Get back to the stern and watch that we don't back over the *Whaler* or its tow line while we're anchoring up."

I hurried to the stern, grabbed the *Whaler's* towline and pulled up the slack so it wouldn't get sucked into the propellers as we backed away from the bow anchor. The *Whaler* also tried its best to get under the stern, and it took a lot of coaxing with a boat hook to push it clear. I was still struggling when the *Swordfish's* engines abruptly fell silent. Ted had shut them down after he had tested the anchor's hold.

After all the aural abuse from the exhaust, my ears still rang in the sudden quiet. The ocean's surface was tabletop smooth, so there was no slapping of waves against the hull. There was instead the noise of water flowing in a stream. The tidal current was extremely strong; even though the *Swordfish* was securely anchored to the bottom, the water rushing past created a small wake as if it were still underway. An occasional call from birds flying overhead added a soprano voice to the now undisturbed chorus of nature. It was the perfect antidote to the diesel engine's previous cacophony.

Ted came aft from the wheelhouse and grabbed a coil of line from out of the cork life raft on top of the main cabin roof. Attached to one end of the rope was a styrofoam ball buoy. He tied the free end to a cleat on the *Swordfish's* stern and heaved the coil into the water behind the boat. The current readily carried away the line and stretched the rope out to its full hundred-foot length. The ball buoy wrestled at the surface, caught between its tether and the force of the tide. "In case someone decides to fall in during the night." Ted answered my quizzical look. "If you fell overboard and the rest of us were sleeping, you'd be half way to Cuba by the time we figured out you were missing. Even with this much line out, the current here is so fast that you'd probably only get one grab at it before you went past the ball. I used to just let the *Whaler* trail further behind us as a safety line, but sometimes a strong wind would come up in the night and push it off to the side of the current stream. The rope floating behind us isn't affected by the wind, and pretty well follows where anyone who might fall in would be carried."

With everything secured, the crew settled around the boat in various places either talking or reading. Ted returned to the confines of the wheelhouse and I heard Wally down in the galley as he made dinner. Smells of the cooking food only added to the pleasant evening. There was still no wind, and all around us cumulus clouds were lit with the vibrant pastel drama of the approaching sunset. The sun's last rays deepened the Marquesas' colors - the white beach contrasted against its rich emerald green mangrove crown. With the extreme clarity of the water I saw the turquoise hued bottom features for some distance around the *Swordfish*.

The sun's face turned orange just above the horizon and focused the last heat of the day in our direction. Most of us had climbed up on the wheelhouse roof to get a better perspective of the sunset. I remembered Ted's story about the Marquesas, how the Spanish had stayed here hundreds of years ago while looking for the *Atocha* and *Margarita*. The beach behind us may well have been the location of their salvage camp. Now it was vacant and serene, as if never trod upon by humans.

Sunset over the distant *Atocha* site, viewed from the Marquesas

Just after the last of the sun had disappeared into the sea, a light cool breeze passed over us. Was this the breath of past history or simply nature marking the coming of night?

The next morning we were underway just after dawn. This time we passed the *Arbutus*, and continued west. Ted hadn't communicated anything to us about his plans for the day. As far as I could tell, he liked to impart information only on a need-to-know basis. I sat quietly with the other divers around the wheelhouse and waited for Ted's directives.

He slowed the *Swordfish* and after studying the LORAN-C's coordinates, looked straight at me and said, "Get your mask, fins and snorkel. Be ready to go at the dive ladder in five minutes."

I sensed that this might be some sort of test so I hurriedly went to the dive locker, found the specified gear and put it on near the ladder on the back deck. Like most of the other divers, I had gotten into the habit of only wearing a swimsuit during working hours, so I wasted no time changing clothes. The *Swordfish* idled, and from its zigzag course I assumed Ted was looking for a certain spot.

The engines reversed briefly to stop the vessel's forward motion, and then went into neutral. Ted leaned out of the wheelhouse door and yelled back to me "We left some big anchors in this area last summer. There should be a large mushroom anchor on the bottom right around here. Snorkel around until you find it. When you see the anchor, stay on the surface right above it and I'll come back and give you a buoy to tie to it. Go now!"

I quickly scanned the horizon, teethed the snorkel mouthpiece and jumped off the side. We were in the middle of nowhere with nothing on the horizon except the distant *Arbutus*. I had never previously swum in the ocean this far offshore. Along with the jolt of hitting the water, brief film segments of shark feeding frenzies played in my memory.

As my entry bubbles dispersed I strained to see the bottom. The water wasn't crystal clear, but the bottom's sandy contours were just visible about twenty feet down. Even with my head underwater I could hear the *Swordfish*, though the sound was noticeably receding. For an anxious minute or two I scanned the underwater vista more for hostile marine life than for the anchor, but there was nothing to see.

I held my breath and made a few free dives to the bottom. Up close it looked like an underwater desert with dunes shaped not by winds but by the tidal flow of the sea. The sand granules were coarse and large; in fact it was not really true rock sand, but flakes of calcium detritus from marine plants and animals. Occasionally broken off blades of sea grass tumbled by in the current. Time here was not measured by clocks, but by the pendulum of the tides.

At the surface I looked again for the *Swordfish* and was dismayed to find it a considerable distance away. I could just make out the form of another diver jumping off at Ted's command and wondered when he would come back to pick me up. Still uncomfortable about treading water alone in the middle of the ocean, I continued my search for the anchor. Perhaps if I found it and could get Ted's attention, he would quickly come back and get me.

During a subsequent free dive, I found a rope, which snaked across the bottom and followed it to the half-buried shank of the anchor. At the surface I tried to hold my position above it against the current while I waved at the *Swordfish*. Finally Ted saw me, turned the boat and idled to my location. A crewmember threw me a buoy and on a breath I tied the loose end of its rope to the anchor. I surfaced near the dive ladder and climbed up to the vibrating deck.

Ted had dropped off several other divers on coordinates of other anchors in the area. As each was found, buoys were also tied to them. Within an hour we had located a total of five.

While we had been engaged in our activities, no one had noticed the approach of the *Jefferson* from the east. Only when she was about a half mile distant did Ted contact her on the radio to acknowledge her presence and describe the position of the anchors we had found.

The plan was for the *Jefferson* to use the anchors as a mooring system for her planned excavation in this area. The deep sand here had been dug previously by the company boats, and some *Atocha* artifacts had been found. Unfortunately, they were widely scattered and found buried under at least twelve feet of sand. Everyone hoped that the *Jefferson's* power would speed up the task of exploratory excavation.

Jefferson sets its anchors in the mist

We watched as the *Jefferson* slowly maneuvered into the area of the buoyed anchors. As the tug pulled up to one of the anchor buoys, one of her divers jumped over the side with the end of a hawser to shackle to the anchor. Once that was finished, the diver climbed back on board and *Jefferson* then moved to the next anchor buoy to repeat the task. Soon all anchors had hawsers attached to them and the tug was like a wheel hub suspended by spokes of winch-tightened lines. She could then lower her deflector and start excavation.

With our *Jefferson* duties completed, Ted pointed the *Swordfish's* bow to the northwest and pushed the throttles forward. After about ten minutes on this course, he slowed the boat to idle and ordered an older diver named Steve and me to put on full dive gear and stand by for his signal to jump. We quickly put on our equipment and assembled at a gap in the railing on the back deck. We had barely settled there when Ted leaned out the wheelhouse doorway and shouted "There's a lot of steel wreckage on the bottom here from the old north tower. There might be another large anchor here as well. Check out the area and let me know what you see. Do it!"

Steve jumped at the command and I followed carrying an aluminum probe for feeling around under the sand. With our entrance, bubbles around us quickly retreated to the surface and we saw the tangle of rusty metal framework about fifteen feet below. I slowly swam downward and noticed that Steve and I were contained in the eye of a silver icthyalogical hurricane.

A cylinder of barracudas formed around us from the sandy bottom all the way to the surface. Hundreds, maybe thousands of individuals formed a shiny revolving curtain around us. As we continued downwards and then along the wreckage, they kept pace with our progress while keeping an equidistant radius around us. The wall of barracudas was so thick that nothing was visible past them. The most barracudas I had ever seen before on a dive were perhaps four or five, this was stunning. I watched Steve closely to see if he was concerned about so many, but he seemed to just enjoy the view. I had never before been in the focus of so many unblinking eyes.

As we swam along the encrusted metal lying on the sand, we came upon a particularly dense area of wreckage. Poised near it were two enormous brown mottled jewfish, each at least five feet long. They watched us approach and held their ground, unintimidated by our presence. I swam slowly up to the nearest one and gave him a gentle tap with the probe right on the forehead. I expected him to bolt at the touch but instead he just continued to scan me with his large swiveling eyes while breathing slowly through his wide, thick-lipped mouth. Steve and I continued our survey of the area, now being followed in dog fashion by both jewfish and barracuda.

The north tower had originally been built years before by Treasure Salvors as a platform for a surveying device called a theodolite. It had been used during the early stages of the search for the *Atocha* along with several other towers for controlling the initial electronic surveys for the shipwreck. Over time these unlighted towers were deemed a hazard to other marine traffic and were pulled down once other methods of navigation came into use. Lots of other junk had also been discarded around the tower wreckage, but we saw no sign of an anchor anywhere. Steve and I slowly rose to the surface towards the thrum of the *Swordfish*, escorted by our cyclone of barracudas.

We climbed up the ladder, reported what we had seen, and molted our dive gear as Ted turned the *Swordfish* in the general direction of the *Jefferson*. As we passed her we saw that she had already put down her propwash deflector and was digging. There was a rolling boil of sand laden water behind her stern; the tide was carrying the plume of discolored water away, much as wind carries smoke from a fire. It was exciting to watch, and the *Swordfish's* divers all hoped that Ted had plans for us to join in excavating the site.

Ted held our cruising speed on a course to the south. When we were about one mile from the *Jefferson*, he pulled back the throttles to neutral and let the boat glide to a stop in the easy swell.

He ordered Wally and another diver to get the "mag fish" ready for deployment. While they worked on the back deck with that equipment, I stayed in the wheelhouse and watched as Ted studied a chart and wrote coordinates in his notebook.

A sudden loud tapping on the back window of the wheelhouse indicated that Wally wanted the window slid open. I complied, and then grabbed the end of an electronic cable he handed me. Ted as usual was totally absorbed in his own work so I just stood there with the cable end in one hand, waiting for directions.

Ted finished with his computations and pulled a scuffed, large black metal box from between the wheelhouse seats. He set the box on a padded shelf adjacent to the back window opening, and then removed the protective casing on the end of the box. The now exposed control panel of the device contained numerous dials, switches and a graph paper recorder. Ted indicated where he wanted me to plug in the connector on the end of my cable. That finished, he also plugged in a power cable connected to two car batteries sitting on the wheelhouse floor.

I had previously read about proton magnetometers and how they worked, but I had never actually seen one until now. They are a very sensitive type of metal detector that registers subtle changes in the earth's local magnetic field, as caused by the presence of ferrous metals. Items such as wrought iron ship's fittings, anchors or cast iron cannonballs can be detected if the towed sensor "fish" passes close enough. Unfortunately, non-ferrous items such as bronze, pewter, lead, silver and gold give no readings at all on this type of equipment.

The "mag" was an imperfect tool, but it had proved itself in locating enough of the ship's iron artifacts to establish a starting point. Excavation near the iron artifacts had produced some silver coins, pottery, cargo items and even a few pieces of gold, but the finds were widely scattered and in very small amounts. The vast majority of the ship still lay undiscovered. If more iron artifacts from the *Atocha* could be found with the "mag", they would give us an indication of where the rest of the ship was.

Ted got the *Swordfish* underway again, with the "mag fish" deployed out about one hundred feet behind us. We proceeded at a fairly slow pace so the heavy "fish" would tow closer to the sea bottom. There was also an oval shaped float on a rope that trailed behind us on the surface. It was tied off at the same distance from the boat as the "fish" for visual reference.

Ted had worked out coordinates for the four corners of his search grid and used the LORAN as his navigational reference. We made advancing passes through the area much like a farmer plowing subsequent furrows in a field. Ted also closely watched the running depth sounder for any rise in the bottom that might snag the "fish."

I watched in silence as one of the more senior divers turned on the "mag" at Ted's nod. He sat facing the "mag's" chart recorder as its needle nervously plotted the local magnetic field. Several concrete blocks with lengths of rope connecting them to colored Styrofoam buoy balls had been placed on the catwalk just outside the port wheelhouse door. Other crewmembers stood there at the ready in case the "mag" gave an interesting reading. They could then throw the block over the side and the floating buoy would mark the spot in the water.

Obviously, the "mag fish" would actually indicate an anomaly one hundred feet behind the boat rather than where the buoy was thrown. To mark that actual spot, we would then turn 180 degrees around and pass by that buoy on a reciprocal course. When the towed oval "mag fish" marker float passed near the buoy ball, another buoy would be thrown. This second buoy would mark the actual anomaly area, and the first buoy could then be retrieved to eliminate confusion.

Even with the doors and windows open in the wheelhouse, at our slow speed little breeze moved through. Most of the crew lingered just outside the doors where it was cooler in the unrestricted air. Ted, now with his shirt off, made slight corrections to our course and glimpsed occasionally at the "mag" chart recorder. I stood behind the operator and watched over his shoulder as he made an occasional adjustment in the settings of the machine. Through several passes in our search area the pen marked slight deviations from its normally straight path, neither Ted nor the operator looked impressed.

After about an hour in the stifling wheelhouse I got pretty bored.

When I looked around I noticed that everyone but Ted seemed to be in about the same state. From a distance we still saw the *Jefferson*, and I imagined her crew diving in the cool water, probably finding tons of artifacts while we trolled the mag. The pen still wobbled its trace along the advancing graph paper and the wheelhouse was a sweatbox. Just about the time I was going to step outside for some air, the operator said, "Here we go...".

Ted watched as well. The pen moved from its position in the center of the paper and accelerated to the right. It nearly reached the edge of the paper before it snapped quickly all the way to the left and then returned slowly to the center again. Just as the pen had shifted to the other side of the paper, Ted yelled, "Throw it!" to the divers who stood next to the waiting concrete block buoys. The immediate noise of a loud splash indicated the buoy had been thrown. The quick turn of events livened the crew like a shot of caffeine.

Ted quickly turned the *Swordfish* toward the reciprocal course so we could mark the actual anomaly. As we came about to this new heading, he spoke to me. "See how the pen went from one side to the other? That means we went directly over the target. If we had passed to one side of the anomaly, the recorder would have also indicated a deviation only to one side before the pen returned to the center. We went straight over the top of whatever it is, so we got both positive and negative sides of the field distortion. It's pretty big too... The pen almost went full scale on both sides. That's the kind of "hit" you want to register."

After we dropped the second buoy on the anomaly, we made several passes near it from different directions. Every time the marker float that towed above the submerged "fish" passed the anomaly buoy, the mag repeated its original reading. We had pinpointed something large made of iron or steel.

Could this "hit" be from the *Atocha?* We were so far from land that to me it seemed unlikely that it could be anything else. Already the adrenaline pumped and the National Geographic theme song played in my head. It surely seemed to me that we might have just found the rest of the ship!

I **HAD** to dive on this spot, but I wasn't quite sure of the proper protocol. As usual, none of the other divers had said anything yet, but gathered in close to the wheelhouse windows and doors for Ted's anticipated next order. I was the new guy, and not likely to be chosen to explore the anomaly. It was time to stand up and be noticed. "Ted, if we're going to dive here, I'd like to be one of the ones to go. I'm familiar with the hand held metal detector - if you'll remember you checked me out on it in port."

Ted replied through his half smile. "A volunteer, huh? OK... Joe - you and Syd go get your gear on. Explain to him how a circle search works. The fathometer shows a lump on the bottom right near the "hit". The water is about forty-five feet deep. By the time you're ready we'll be anchored up. The rest of you guys, get the anchor ready - and Wally, watch the *Whaler* when we back up."

While the *Swordfish* anchored near the anomaly buoy, Joe and I assembled our equipment on the back deck. As we put it on he explained the circle search procedure.

We would both descend down the buoy line to the bottom, where he would tie one end of a rope marked in five-foot increments to the concrete block. I would turn on the hand held metal detector and tune it. He would then let out the first five feet of rope from the concrete block, and keeping light tension on it would swim the complete circle described by that radius. I would be following him, sweeping the search coil back and forth between the concrete block and the five-foot mark in his hand. When we got back to our starting point, he would let out five more feet and we would continue the next circle. We would continue adding concentric circles until we found the "hit".

We used the hand held metal detector because it was likely that the metal object for which we were looking was buried. The one we had on board was the only one that the company owned. Most of it was encased in bits of PVC pipe that was glued and clamped together in a fashion that shouted: HOMEMADE! Only the large black search coil on one end looked professionally built. Just picking up the device caused it to creak and sag greatly from the stress of supporting its own weight. Jumping in the water with it was a big no-no, unless you enjoyed looking for scattered parts.

The engines stopped, indicating we were anchored. After Joe got the nod from Ted, he jumped in the water and swam toward the buoy that floated about twenty feet off the port side. I grabbed the detector and gently climbed down the dive ladder into the water. By the time I got to the buoy, Joe had already gone down. I followed in his bubble trail toward the bottom.

When I arrived, Joe knelt next to the concrete block and tied his search rope to it. The visibility was only about eight feet and the water was filled with tiny motes of silt that moved slowly in the tide. It was a greenish-grey world with no direct light or shadows. I turned on the detector, put the earphone under my mask strap and with the coil on the bottom adjusted the sensitivity knob.

As we began the search I had to be careful not to stir up the soft bottom with my fins or the detector coil. By the time we finished the first circle though, visibility had dropped to about three feet. On the second circle I noticed at one point a faintly darker form just at the limit of our visibility. The third circle brought us close enough to the object for identification. It was a small, dead, silted-over patch reef.

I moved the metal detector's coil towards the edge of the reef. A tone sounded through the earphone and quickly rose in pitch. Joe could hear the noise of the earphone too, and turned to see what I had found. The detector seemed to indicate a strong reading in the silt immediately adjacent to the base of the patch reef. Even with the coil turned on its side for minimum sensitivity, I could still pick up whatever it was over an area about five feet long.

I indicated to Joe where I thought the center of the "hit" was and he came over and joined me in digging in the mud with our hands. Neither of us had thought to bring any kind of digging tool and the broken fragments of shell in the mud were painful to our fingers. Joe stopped digging and through pantomime indicated that he was going to the surface to get something with which to dig. He disappeared and I kept working.

About eight inches down I could feel something hard. The visibility was terrible from the displaced silt, so I had to explore the object completely by feel. There were sharp shells and barnacles attached to whatever it was, but the overall contour seemed too perfect for a natural formation. I continued to clear away the mud.

When I had exposed about one square foot of it, I stopped digging and let the current take away some of the turbidity. Slowly the murk shifted away and I could see a shape. It appeared to be cylindrical, indicating a width of perhaps sixteen inches. With the butt end of my dive knife I tapped on the object. It seemed quite hard. I really banged on it, and some of the shelly concretion flaked off to show a dark metal underneath.

Now I was excited. It had to be cannon! I knew that in 1975, nine cannons had been found not far from other scattered artifacts from the *Atocha*. We knew from Dr. Eugene Lyon's research that the ship carried eighteen on board, so it was already obvious to me that this was one of the missing guns! Here I was on my first trip and I had already found a major piece of the wreck - the new guy really shows 'em how it's done! I quickly checked my air supply and then began digging furiously towards one end of the cylinder. I'd have it all dug out by the time Joe got back.

The visibility again went to zero, so to occupy my time while it cleared I banged the freshly exposed area with my knife again to knock off the encrustation. Maybe there was some engraving or design cast onto the gun that would be visible with the encrustation removed. What was keeping Joe?

The tide gradually drew aside the curtain of silt. I paused and looked closely at the exposed metal to see if any detail had been exposed. With more of the object out of the mud I saw something I had not noticed before. Fins now protruded from the exposed end. It was not a cannon... it was a large, intact bomb.

Of course. I had once heard Ted talking about how since World War II the military had used this area as a bombing range. A twentieth century weapon instead of a seventeenth century one. How long does it take for one of these things to become inert?

I instinctively backed away just as Joe returned carrying the shovel out of the *Swordfish's* icebox. He looked at the bomb, shrugged and began coiling up the circle search rope as he swam back to the concrete block. His actions confirmed that we had found the "hit".

My inexperience hurt more than the cuts on my fingers. In the excitement I had forgotten that the *Atocha* carried only bronze cannons, not iron. In her day the metallurgy to produce reliable cast iron guns had yet to be developed. Nothing made of iron and shaped like this would have been on board. My ego-cannon fantasy had pushed me to pound a hard object on an intact five hundred-pound bomb rather than explore more carefully. If it had actually been an artifact from the *Atocha*, I might very well have damaged it.

As we swam to the surface I resolved to think first next time before I acted. I hoped to never see a bomb again, or at least to be able to recognize one quickly for what it was. I didn't know then that I would get plenty of practice. We would find thousands.

Company Man

During the following weeks the *Swordfish* crew cycled between *Galleon* repairs in port, and short mag survey trips to the *Atocha* site. The paychecks stayed irregular, but since we were going to sea regularly there was just enough food on board to get by. For now, going out to the site was entertaining enough that a steady income didn't seem all that important.

That was a good thing, since none of the divers really understood how fragile Treasure Salvors' finances really were. The American economy was in the middle of a serious recession. Mel did his best, but finding fresh investors to keep the expedition going was tough. Ted quietly began to finance much of the operation himself. We didn't know it, but he had stopped drawing his own salary so the divers were able to get something in their pay envelopes every week. Money from his own retirement savings bought equipment and fuel for the boats. Mel promised to pay him back when things got better, but they both knew there were no guarantees.

Treasure Salvors needed to make some new finds on the site. If something newsworthy was recovered, finding investors would be easier. One option was to have the *Jefferson* excavate the area that had previously produced artifacts, hoping that her exceptional digging capabilities would turn up something new. The other was to expand the search for the missing bulk of the *Atocha* by expanding our mag surveys. We would do both as long as the money held out.

The *Swordfish* spent long days continuing its survey of the **Quicksands**. I learned much more about operating the mag, but would rather have had the chance to dive in the *Jefferson's* excavations. From radio conversations, we knew that the *Jefferson's* divers had been finding some *Atocha* artifacts.

As important as it was, magging was mostly boring. So far every "hit" we had found and dove on had turned out to be either bombs or other modern junk. The quirky, tired mag sometimes gave false readings or failed altogether. We longed to find anything that related to the *Atocha*, which was made harder when the senior crew members described how much fun it was to work in underwater excavations. "Mag is a drag...digging's a blast!" one of them chanted.

The time had come for a crew rotation aboard the *Jefferson*. Ted stated that I now had enough experience to take my turn aboard. The normal stay aboard for our divers was roughly two weeks. Since we were currently back in Key West, the crew change would take place when the *Jefferson* returned at the end of her current trip.

One morning Ted ordered me to get my gear together and get down to pier "A" where the *Jefferson* had just docked. She had returned in Key West to refuel, reprovision and fix a mechanical problem. The difficulty could be fixed in a day's time, and we would then head directly back out to the site.

Mel Fisher attempting to convince *Galleon* museum visitors to become investors in the *Atocha* project by displaying some early finds

Happily for me, Jeff would also be on board. He had several trip rotations already under his belt and was one week into his current turn. I looked forward to his insights on what was expected of us. Surprisingly, Fred still kept his bunk on board as well. Though the *Jeffersron* only needed two divers, he was getting as difficult to remove from the boat as a barnacle.

Since all the bunks in the wood paneled staterooms were full, my sleeping quarters were literally a step down. Taking a steep flight of stairs below from the main bunkroom, I went forward to my assigned bunk space adjacent to the bow rope locker. It was a stark, grey and white painted metal room contoured by the shape of the hull. It felt a bit like being in a submarine since it was well below the waterline and had no windows. The only light was a single bulb that glared from its vapor proof housing. Noise from the waves expending themselves against the metal hull was as noticeable as the smell of fresh paint. The bed was alone, a steel frame welded to the deck in the center of the compartment.

Several years earlier, one of Mel's salvage boats called the *Northwind* had rolled over and sunk in the night while anchored near the wreck site. Though most of the crew survived the accident, Mel and Deo's own son Dirk, his wife Angel and diver Rick Gage were trapped inside and drowned. The circumstances of the tragedy had been a poignant element in the National Geographic documentary on the *Atocha* search. The incident was discussed amongst the current crew only around newly hired employees, and almost as a mythological story. With the ocean's perennial voice against the hull of the *Jefferson* and *with* the *Northland's* sad end in mind, I made a point to keep a full scuba tank and regulator under my bunk.

Jeff had been released from any duties onboard while the *Jefferson's* engine was being repaired, so he was eager to show me other points of interest on board. One area I had not seen before on our previous visit was the "head" – the nautical term for a ship's bathroom. After my query, he led the way saying, "The fountain of youth? You're gonna love this!"

We walked outside to the main deck and around the deckhouse to the port side. We arrived at a closed watertight door. He grabbed the handle to open it and said, "Maybe I should have had you bring your dive mask..."

As I opened the door I looked into a small, white painted compartment complete with sink, mirror, toilet and a small shower area. Every surface, including the inside of the door and ceiling, dripped with water. The roll of toilet paper sitting on the sink was a sodden paper mache object d'art. Jeff stepped back slightly. "Check out the bowl, man."

With every wave that washed against the side of the tug, the water inside the toilet bowl welled up violently, almost slapping the bottom of the seat. Since not all the waves that lapped against the hull were the same size, the disturbance was not regular in amplitude. As a fishing boat made its way up the channel past the *Jefferson's* pier, Jeff grabbed my arm and pulled me back. "Perfect timing - watch this."

When the fishing boat's wake rolled into the side of the tug, the water in the toilet bowl exploded upward as if driven by a fire hose. The stream hit the ceiling and was deflected in all directions, even outside though the door and onto us. Jeff laughed. "I'm not sure whether this door is here to keep water in or out of the head. If you're going to use this thing, you better be quick and have good timing. The first time I used it, I was sitting there nice and comfortable with a magazine… Next thing I know I'm soaking wet covered in shit. It was in the sink, in my hair and all over the floor. I had just taken a shower and put on fresh clothes. Good thing they put large drain holes in the floor, I had to hose the place out. It's probably the best bidet in the Keys!"

Boat heads never work as well as those on shore, but I'd never seen one do this. The discharge pipe came out of the hull just at sea level and somehow the system allowed water to go back up the plumbing every time a wave smacked it. Out at sea in unprotected waters it would be quite a show. Maybe the *Jefferson* wasn't so posh after all.

The next day we got underway just after dawn. The owner's son, Danny was captain and seemed quite knowledgeable in handling his vessel. Although Danny had much seagoing tug experience, he had never previously been involved with anything like the *Atocha* project. The tug's operation was completely dependant on site information from Treasure Salvors and feedback from the divers. Neither he nor his first mate dove; they were there just to put the *Jefferson* where the company told them and to support the diver's efforts.

On the way out to the site, Jeff and I lounged atop the deckhouse. It was noisy since the tug's exhaust stack discharged just above, but it was pleasantly sunny and away from Fred who would rarely leave the air conditioning. For some reason he was still onboard, overly polite and attentive around First Mate John and Capt. Danny, gruff and officious around us.

In a recent alcohol-prompted confession, Fred stated that he hoped to find a permanent position as a *Jefferson* crewmember. His eagerness to impress John and Danny was pretty transparent to us but they seemed to accept him at face value. He had apparently evolved his position onboard to a kind of intermediary between the management on the tug and the *Treasure Salvor* divers. In other words he made tropical drinks on demand for Danny and John while sidestepping any physical labor by giving those types of projects to others.

There was a large, covered fiberglass tub mounted on top the deckhouse that served as a temporary storage container for artifacts recovered by the *Jefferson*. All items recovered on the site were immediately tagged and placed in the tub as soon as the divers brought them to the surface. The tub was filled with fresh water, as the artifacts would rapidly deteriorate upon exposure to air. Normally, when the *Jefferson* returned to port, the items would be off loaded and transferred to Treasure Salvor's conservation lab where the stabilization process began. Since the tug had such a quick turn around this time there hadn't been opportunity to take them ashore.

Anxious to show me the things they had recovered during the first part of his trip, Jeff fished several items out of the tank. "Here's a pottery shard I found on my first dive. There were several ballast stones in the same excavation so I was really gettin' excited…it looked like a pretty good area. Joe was in the same 'hole' with me and he had the metal detector. He found this E.O. which looks like either a small spike or maybe part of a barrel hoop."

The ballast stones Jeff mentioned were a common method of ballasting ships in the 17th century. They were mostly erosion-smoothed rocks such as those found in riverbeds, varying in weight from five to about forty pounds. They were placed in the bottom of ship's hulls to counteract the wind's tipping effect on the sails and tall masts, much like a modern sailboat having a weighted keel. An "E.O." was company slang for "encrusted object." In many cases the actual shape of an artifact was masked by a thick coating of corrosion byproduct and marine encrustation. Any item, which could not be readily identified upon recovery, was logged as an "E.O.".

Jeff saved the best for last. "The next hole was my turn with the detector. As soon as we stopped digging, I went over the side almost before the visibility cleared. I had the detector turned on before I hit the bottom, and as I worked toward the back of the excavation I got a nice 'hit' on it. Check it out!"

In his hand he held an irregularly edged flat disc. Blackened with silver chlorides and sulfides, as a result of centuries of salt-water immersion was a silver coin. A famous Spanish piece-of-eight.

I was thrilled and immediately jealous. Though finding anything from the *Atocha* would be exciting, a coin really was something special. Not that many had been found so far and few of the divers currently at Treasure Salvors had actually found one. This was Jeff's first coin and the only one that had been found so far aboard the *Jefferson*.

He held it out toward me as he turned it with his fingers so I could get a better look. It slipped from his hand, bounced edge-on right off the cabin roof deck and flipped off the edge toward the lower deck. We watched in silence, as it bounced again on the flat gunwale, seemed to pause briefly in mid-air and then disappeared into the flow of water alongside the *Jefferson's* hull.

Oops…

The mood aboard the tug was a lot less exuberant after everyone was told about the coin's loss. It had symbolized the first real success of the *Jefferson's* efforts. We would just have to find some more, though so far the rate of artifact recovery had been pretty slow.

On November 2nd, when we finally arrived on site, Danny maneuvered the tug through the field of buoys that marked his last excavations and where the ship's anchors had been left on the bottom. As he nosed the tug up to each anchor buoy, either Jeff or I jumped into the water with full scuba gear, taking an end of one of the anchor hawsers and shackling it to a half-buried anchor. The *Jefferson* then moved quickly to the next buoy while the deck crew paid out slack on the hawser that had just been shackled. Four anchors defining the shape of a very large rectangle around the tug were eventually hooked up. The task of guiding the vessel against the strong tide while trailing each additional hawser within the boundaries of the anchors was quite difficult. The slightest mistake could cause the *Jefferson* to end up hopelessly twisted and fouled in its own ropes, held by the tide like an ant in a spiderweb.

The deck crew began winching in on the various anchor lines to position the boat in the right spot. Jeff and I climbed back aboard up the dive ladder, both thoroughly winded by our underwater exertions. Before the hour was out the vessel was in the desired location and the propwash deflector was lowered and secured. That accomplished, the *Jefferson's* locomotive engine exhaled hard, driving seawater against the deflector onto the ocean floor below.

As Jeff and I switched to full tanks, we watched the sand displaced by the propeller's force roil the surface behind the stern. The ocean's blue-green translucent complexion was marred by the growing khaki bloom of disturbed sediment carried away by the tide. The sand was thought to be about ten feet deep here, and it would take roughly ten minutes of digging to make a worthy excavation. While preparing our gear, we discussed the plan for examining the excavation when it was completed.

Because of his experience, Jeff would operate our usual creaky metal detector and I would visually examine the hole while he electronically surveyed it. Ted had loaned Treasure Salvors' only detector to the *Jefferson* for our use on this trip. We handled it learily, knowing that if it broke the whole operation would come to an abrupt end. I was just as happy to let Jeff have the responsibility of its care.

The tug's loud exhaust suddenly dropped 10 decibels as Danny pulled the throttle back to idle. For two minutes the propeller turned at this power setting and then he shut down the main engine. With just the muffled rattle of the ship's generator left to spoil the natural sounds of the sea, we adjusted our dive gear and waited for the all clear from Capt. Danny before jumping in.

In a moment he appeared through the wheelhouse door giving us a thumbs up. First Mate John stepped on deck from the engine room in time to mock-stutter to us; "Go-Go-Go Gold !" I jumped over the side while Jeff teetered down the dive ladder clutching the fragile detector.

The tide quickly carried the last of the silt-laden water away as I swam down. I was determined to get to the bottom first in case anything had been uncovered that could be seen by human eye. In the sea floor not far from the stern of the tug began a large trench that looked much like an elongated bomb crater. The sand had been piled in a huge berm all around the excavation but especially so at the end farthest from the boat. Ivory white flat bedrock lay exposed for the length of the excavation. Down near the bedrock the view up the steep sand berm was quite impressive. The *Jefferson's* power had moved many tons of sand and had in just a few minutes cleared a patch of bedrock twenty by eighty feet.

As I swam along the boundary of the sand berm and the bedrock, I saw nothing remarkable. Jeff arrived on scene and paused in the middle of the cleared seafloor to tune the detector before beginning his sweeps. I knew from previous discussions with him that most of the artifacts found seemed to be at the part of the hole furthest from the boat, so I moved in that direction.

The sand and bedrock features were all a light monochromatic tan. The flat bedrock was pocked with thousands of holes of various sizes from near microscopic to as large as my head. I could see something of a darker hue ahead as I swam toward the back of the excavation.

Several dark grey granite ballast stones lay exposed on the bedrock

Recovered *Atocha* ballast stones

next to the sand interface. They weren't particularly large ones… perhaps five pounds apiece, but they had the telltale shape of erosion-molded shoreline rocks. The bedrock and sand on the seafloor here were all produced as a by-product of plants and animals. These black rocks were as foreign to this locale as I was.

I picked up one. These were actual artifacts from the *Atocha* on the bottom of the sea. I was the first person in nearly four centuries to touch them. They had been hand placed in the *Atocha's* lower hull as ballast to augment the weight of the cargo. Though only rocks, at that moment of discovery, they might as well have been solid gold. Here I was, on the ocean floor looking at artifacts from a real Spanish galleon. My dream was coming true.

Jeff in the meantime had been working the excavation with his detector. Several readings on the machine produced a few heavily encrusted barrel hoop fragments from the *Atocha* along with modern rusty pieces of steel and slivers of corroded aluminum cans. As soon as we finished examining the entirety of the excavation, we kicked out of the sand walled canyon up towards the *Jefferson*.

Back on board, we reported our finds and their positions in the excavation to Danny and John, who logged the coordinates, tagged each artifact and placed it in the storage tank. The modern junk was thrown in a garbage can on the back deck.

Over the next few days this digging/diving routine was repeated many dozens of times. From the dawn's pale blush to evening's dim withdrawal, we winched ourselves along what we believed to be the *Atocha's* trail of destruction. For our efforts we discovered more ballast stones, barrel hoop fragments and even a few small pieces of broken pottery. The tally of modern trash to be found was much greater and certainly more varied. It was heartbreaking to spend 20 minutes trying to dig out a detector "hit" from the collapsing sand berm wall only to finally produce a beer can tab or a rusty bolt or mattress spring. The sands west of the Marquesas Keys held all manner of objects from human endeavor and gave them up with the same indifference.

The long workdays were tiring, and in addition we were all expected to keep watch at night. The watch duties included inspecting the bilge spaces for flooding, checking the frequency of the generator, watching for vessels or storms passing through the area and looking for problems with the anchors or their lines. Naturally, being the new guy entitled me to the watch schedule no one else wanted - from one to five o'clock in the morning.

After relieving Jeff at one o'clock in the morning, I would wander around with my flashlight, moving just enough to stay awake. After fulfilling my responsibilities for the moment I often stepped outside and divided my attention between the two night universes of sea and sky. Both were pitch black;

the one above flung with more stars than I had ever seen and the one below trembled with the phosphorescent discourse of survival. The ocean was alive at night.

After more than a week had passed on the site, Capt. Danny made the decision to head back to Key West for more provisions and fuel. We would spend our final morning recovering anchor lines from the anchors. This operation should have been the reverse of our mooring setup on the first day.

In the middle of our efforts to unshackle the hawsers, a sudden, powerful storm moved through the area. This caused the tug to twist in its remaining moorings and drag sideways through the anchorage, snagging buoys and lines as it went. Danny tried to use the engine to steer us out of the tangle but we ended up abeam to the seas with a hawser and anchor crown buoy line wrapped around the prop. Six-foot waves smashed against the side of the *Jefferson*, flooding the low-lying back deck with several feet of seawater. The scuppers (drains) in the gunwales (sides of the boat above the deck) could not drain the water as fast as it was coming aboard. Already floating coils of rope and hawser were in danger of being carried over the side. Everyone stood in water up to his knees.

Jeff, Fred, Capt. Danny and John worked the two winches and tried to grab floating equipment. I still had my dive gear on so John handed me a knife and yelled over the din "You've got to cut the lines out of the prop. These last two anchors we're hangin' on are draggin' - we are trying to hold our position by winching to them but we've got to use the engine to get out of this mess. Be careful!"

I put on my mask, bit on the regulator mouthpiece and jumped over the lee side. In addition to the waves and wind that tried to push the boat, the tide also ran hard in the same direction, carrying me away. I swam quickly down to the bottom, crawling and kicking my way back to the tug. From the moment I had entered the water I was aware of the wave's rock hammer blows against the *Jefferson's* steel side. Approaching the bottom of the vessel was physically painful; the air-filled voids of my lungs, sinuses, and ears wrenched from each concussion.

I saw both the thick hawser and the anchor's crown buoy line wrapped and looped tightly over the prop blades. The hull of the tug rose about five feet with each approaching swell and then violently slammed down after the booming wave strike. This was not going to be easy.

I timed my approach to the prop so that I made contact at the bottom of the swell. As the tug's hull began to rise I cut furiously at the ropes. Just before the approaching wave crest hit, I paused and grabbed the looming propeller blade just above my head. Simultaneously came the thunderclap of sound/pain as 150 tons of tugboat dropped down hard on me. Then with the next rise I started to cut again.

The unforgiving prop blade tried to chop down hard on my head each time the tug plunged deeply into a trough. This was a very unforgiving situation; if I were incapacitated no one on deck would know. The ropes I sliced at were alternately slack and then pulled guitar string tight from the oscillation of the hull. It would be easy to get my fingers caught at the wrong moment. After I sawed for nearly ten minutes, the last fragments broke free.

I quickly surfaced and pulled myself up the dive ladder, trying to catch my breath on the still flooded aft deck. John motioned for me to help gather up the hawser that was floating around the base of the winch on which he was working, and keep it from fouling him or the winch. I shucked my dive gear into the knee-deep water and waded over to help.

Capt. Danny, up on the wheelhouse catwalk directed Jeff and Fred on the bow winch. John on the aft deck winch tried to move the tug sideways against the seas so we wouldn't get caught again in more awash lines that were just off our leeward side. If we could just pull another twenty feet against the seas, we could engage the engine and move forward quickly before the stern of the tug fell upon the rest of the half floating anchor lines.

John's winch was really getting dangerous; the force of the current, gale force winds and the increasing waves multiplied the load on it. The rope wrapped many times around the winch head would sometimes slip back violently, the tension pulling his hands sharply up to the revolving steel drum.

As he worked the winch and I managed the floating line that had been drawn in by it, we both fought the confused flush of water and spray from waves, which thundered over the side. Only the very tops of the gunwales were above water on the back deck. Suddenly the rope on the winch slipped out again and this time John's hand was drawn sharply into the winch.

"Damn" he shouted, grabbing the injured forearm with his other hand. "Take the winch, I think my arm's busted."

The pain evident on his face showed he wasn't exaggerating. I tenuously grabbed the line from the winch drum, and pulled tension on it so the friction would cause the rope coils to bite into the drum's surface. The line once again started to pull super-tight and I tried to make sure my hands were as far away as possible in case it suddenly slipped back on me. This was not a good feeling.

I winched in another 15 feet of rope before Capt. Danny decided that was sufficient. Quickly he pushed the main engine throttle forward and signaled for me to release the line from the winch. We just missed the rest of the mooring and buoy lines and the *Jefferson* quickly swung into the seas on the bow anchor. With the waves rolling alongside the hull instead of breaking over it, the seawater began draining off the back deck. We could finally attend to John's injury and the mess of lines and equipment on the back deck.

Between the winch incident and cutting the lines out of the *Jefferson's* propeller it had been quite a morning. I had just gotten a small taste of the potential for physical harm while out at sea. Being under thirty years old, I was still clad in a self-perceived cloak of invincibility. Regardless, today's lessons had been clear: Pay attention, this is a serious business.

As we stabilized John's arm and made final preparations to get underway to Key West, Danny received a message on the tug's radio. Jeff broke the news to me.

"Maurice on the *Arbutus* called, he wants to go into Key West with us. He has some things he needs to take care of or something. Guess it's your turn to relieve him. Don't be too bummed… Danny says he plans to come back out in a day or two. We'll drop off Maurice and pick you up as soon as we get back. Just take some clothes. Leave the rest of your stuff next to your bunk. I'll keep an eye on it. Besides, anything you take onboard the *Arbutus* is going to get nasty."

I stowed my dive gear bag and anything else of value under my bunk and tossed some clothes in a plastic garbage bag. Minute's later Jeff, Fred and I got in the *Jefferson's* leaky inflatable boat and skidded across the wind-driven seas toward the *Arbutus*.

We had to make the trip quickly since the inflatable had already had one too many encounters with the *Arbutus'* barnacle wreathed hull. Fresh but unsuccessful patches on the boat's rubber side fizzed with escaping air. Adding to the excitement of the deflating boat was its outboard motor's inability to run on both cylinders. Only when we held the throttle wide open did the motor provide enough power to move the boat forward.

Pulling alongside the *Arbutus* was the next challenge. Maurice tossed a thick, knotted rope over the gunwale amidships. Normally the deck was about eight feet above the water but with six-foot waves in fast parade along the ship's hull, the distance varied dramatically from crest to trough. I tossed my garbage bag luggage up to Maurice on one crest and scrambled up rope on the second while Jeff and Fred fought to hold the inflatable in position and away from the *Arbutus'* abrasive side. I was pretty self-impressed with my own ability to make the transfer quickly until I turned around and found that Maurice had already loaded his baggage and himself into the small boat while I had been climbing up. Above the outboard's spastic knocking, Maurice yelled, "Don't use the reading light if you want to have the VHF radio. The battery's almost dead!"

The inflatable had lost much of its rigidity due to the air leaks and as it pulled away from the *Arbutus* the sides seemed to fold up around the passengers. Somehow they made it back to the *Jefferson* successfully, and quickly hauled the cripple back on board. The *Jefferson* soon left the site and began the four-hour trip back to Key West.

I went up to the *Arbutus'* bridge to check the ship's radio operation. The radio itself was small, old and was hooked up to a car battery. Maurice had set up the radio next to the smelly mattress where two loose wires also lead to a 12 volt light fixture on the ceiling right above the bunk. Hooking the battery up to the light required that I disconnect the radio. I turned on the radio, called the *Jefferson* on the appropriate channel and was pleased to hear Capt. Danny respond clearly. He confirmed again that he would return in a day or two. I wished him a safe trip, turned off the radio to conserve the battery and went off to explore my new surroundings.

On my earlier visit to the *Arbutus*, I hadn't had the time to go forward of the bridge area. There was a large open foredeck with steel hatch covers over the entrance to the cargo hold. The winds had moderated considerably since the morning and though the waves were much smaller, the ship still moved about. As I crossed the open foredeck, I noticed a slow but considerable rolling motion. One of the four hatch covers was slightly askew, so I kneeled down and peered through the gap.

The void below was so dark that I couldn't see much, but there was an alarming noise of moving water that heightened with each roll of the ship. There were other strange sounds as well. This certainly required a better look, but the heavy hatch was more than I could move. There had to be another way into the hold area.

At the front of the bow was a large hooded enclosure, which was open towards the foredeck. The ship's bow winch and other equipment were mounted there. To the side of the winch area was a bulkhead with a watertight doorway that faced aft. I entered the door and followed the steps downward.

The stairway emptied into a large room that must have been the enlisted men's berthing area. The noise of moving water was much louder here and I followed the sound through the dim light. The bulkhead at the back of this compartment had a closed watertight door… here the sound was loudest. I timidly moved the "dogs" (handles that secured the door shut) and pulled open the door.

Shafts of light from the askew overhead cargo hatch illuminated the half flooded hold. The deck of this compartment was apparently another level down, indicated by another descending stairway from the doorway in which I stood. Not only was the area flooded but it writhed with unsettled, tumbling junk of all descriptions. Large sections of PVC pipe, old bed frames, scaffolding, oil drums and lots of unidentifiable rusty metal forms were carried back and forth across the hold by the force of the wave wash. I couldn't discern the depth of the water, but the top of another open watertight door visible on the other side of the room indicated that the depth was at least six feet. Was this tub sinking?

I closed and dogged the watertight door and hurriedly made my way back up on deck. The sun was starting to get close to the horizon and the *Jefferson* was well out of sight. There I was, forty miles from Key West on a derelict vessel that seemed to be taking on water. There was no lifeboat, and the only floatation device that I had seen was an old greasy life vest thrown into the corner of the galley. I ran up the stairs to the bridge, turned on the radio and called the *Jefferson* in the calmest voice I could summon.

Capt. Danny answered. "What can we do for ya? Over"

"I just found out the cargo hold is half full of water. I'm not sure of its source, but it looks pretty bad. Maybe this thing sprung a leak. Over"

"Hang on *Arbutus*; Let me talk to Maurice ... stand by."

Several minutes passed. Maurice finally answered the radio. "*Arbutus*, this is *Jefferson*. Ah... I haven't pumped the bilges in a while. There's definitely a leak there somewhere but I gave up tryin' to find it. It's been almost three or four weeks since I last pumped it out, so maybe what you're seein' is accumulation over that time. I wouldn't worry about it, but you can always pump it out if you want to. Sure seems like a lot of water though. Over."

I told Maurice that I definitely wanted to pump it out so he explained the procedure. I was to look for a small gasoline powered pump and the hoses with which to hook it. The equipment should be somewhere on the foredeck. He wished me luck and told me they would be in Key West in about an hour, at which point their radio would no longer reach me.

The sun was rapidly setting. I found the pump lying near some old steel drums. Like its surroundings, it appeared to be made mostly of rust. Weather-worn intake and discharge hoses laid nearby. In short order I positioned the pump near the partially opened hatch cover with the suction end of the hose dangling down into the hold. Amazingly, the pump started without much drama but keeping it primed took a lot of cursing on my part. Finally, the discharge hose kicked as the pump took its draw and spat a powerful stream of bilge water over the side.

I found some extra cans of gas inside a compartment just forward of the galley, so I knew there was enough fuel to run the pump for quite some time. Satisfied with the pumping operation, I retired back to the bridge for a peanut butter sandwich made from ingredients requisitioned from the *Jefferson*. An hour later, my flashlight's limited squint into the hold from the cargo hatch opening showed the water level had indeed dropped. With the sun fully down, the only other light on board was a dim yellow blinking traffic hazard light (long ago liberated from a Key West construction site) that was mounted on top of the wheelhouse as an ersatz anchor light.

Back on the bridge, I lay in Maurice's nasty bunk and read the old magazines he had left by dimming flashlight. The night had turned serene, the velvet sea air barely moved through the open semicircle of wheelhouse windows at the front of the bridge. The pump motor eventually ran out of gas, but I had begun wading through the shallows of sleep.

The sky and the ocean talked to each other. The conversation was at first soft and unintelligible, and as I slept it made no impression upon me. After some time an argument began between the two and in crescendo their voices changed to a screaming rage.

BANG!!! The loudest noise I had ever heard accompanied a light so bright it hurt my eyes through my closed eyelids. Shocked by the blast into a sitting position and still stupid with sleep, I tried to figure out what was happening. A bitter cold wind drove through the open windows and carried with it buckets of sea spray. Blue–white lightning with instant thunder-crack retort was all around. The *Arbutus'* towering steel mast or the bridge itself had been struck by a lightening bolt.

Everything in the bridge was soaked from the driving rain and sea spray that came through the front windows. Jumping to the deck, I navigated the room easily by the light of the almost constant lightning. I tried one window after another and could not budge any of them; they were all seized open by corrosion and time. Outside in the gale, huge waves broke over the sides of the ship as it writhed like a snake in fire. The ship's unbalanced thrashing had become extreme enough that I could not stand without hanging on to something. The bunk was in the area furthest from the gaping windows, so I retreated back to it across the slick floor and climbed back in, pulling the soaked bed sheets over me. This was the most violent weather I had ever witnessed and there was nothing I could do but ride it out. I turned on the VHF radio and tuned it to the weather channel to listen for any special marine warnings but only received static. The rest of the night I shivered under the cold, wet covers.

With the gray dawn, the rain finally ceased but the wind did not. The first order of business for me was to make my quarters more livable. As I scrounged around in the various compartments on the main deck, the only potential lubricant I found was a bottle of oil and vinegar salad dressing. I poured it onto the stuck window tracks, which enabled me to finally slide them shut. Since the glass windows in both bridge access doors were completely missing, I completed my weatherproofing by bolting old pieces of plywood over the openings. With no more spray coming through the windows, my soaked belongings slowly began to dry.

I refilled the bilge pump with gas and restarted it. Next was to figure out why the radio wasn't picking up anything. I followed the antennae cable from the bridge to the main mast and found the cause of the radio's malady.

The antennae and its mount had been blasted off the steel mast by the lightning strike and had fallen to the bridge roof. Surprisingly the antennae looked OK but the mount was fried. Some rope would suffice for tying it back to its perch.

I knew that a VHF radio's range was pretty much line-of-sight. That ment putting the antennae up as high on the mast as possible to allow me to communicate with the *Galleon* office or any other receiver in Key West. This didn't look too daunting, so I climbed the rusted rungs of the sixty-foot tall mast. About twenty-five feet up I began to question the wisdom of the idea. The ship's roll was amplified progressively with each step upward and I was whipped back and forth over the unforgiving steelworks below. If I fell and was injured…well, no one would know until they came back out to the site. I inched slowly up another ten feet and tied the antennae off, hoping it was high enough.

My efforts were rewarded with a successful radio call to the *Galleon*. Ted was actually calling me just as I turned on the radio receiver to test it. He said that a powerful cold front had passed through the area last night and that it was going to be windy for several days. The *Jefferson* would not return for me until the winds had calmed. In the meantime I should enjoy my "All expense paid Caribbean cruise". He also told me to have the radio on at eight in the morning and five in the evening so he could check on me.

The rest of the gloomy day I pumped out the hold area. In between checking out the pump, I further explored the empty steel caverns of *Arbutus'* below deck areas. By nightfall the hold was finally pumped dry. Satisfied that no major leaks existed, I retired to the now snug bridge.

The next week settled into an unfortunate routine. Every day I awoke to find no change in the wind or waves. Ted made his morning radio call to announce more forecasts of continued high winds. I shuffled down to the galley and removed a rusty labeless can from the doorless pantry, opened it with a knife and consumed whatever the contents held. The fresh food that I had brought with me from the *Jefferson* had quickly run out. I subsisted on whatever food Maurice had rejected. Lots of the cans turned out to be beets or green beans.

There was a scuba tank refilling system on board, the only one that Treasure Salvors owned. Like the rest of the company's equipment, it teetered on the brink of junk. The wheezing compressor could be used to fill a series of large high-pressure cylinders, which in turn could be used to fill perhaps ten scuba tanks at one sitting. It took many hours for the compressor to fill all the cylinders but it gave me something to do for a day.

Except for a few gallons of drinking water there was no fresh water aboard. After a few days of not bathing, I, like Maurice, began to take on the same ochre of the *Arbutus*. Finally I gave up and used a bucket on a rope to get some seawater for bathing. I felt cleaner after my dish soap and saltwater scrubbing, but my skin itched constantly from the dried salt. With nothing really constructive to do, each day seemed to last at least fifty hours.

Going into my second week, I really started getting depressed. Stories told amongst the divers back in Key West about the *Arbutus* being haunted or about the ghosts of the *Atocha* were things about which I avoided thinking. They must have been near the surface of my subconscious though, waiting for the right cue.

Late one night I was awakened sharply by the sound of someone running up the steel bridge steps just outside the door next to my bunk. I sat bolt upright and groped the bunk area for my glasses while I stared wide-eyed in the direction of the door and expected someone or something to burst in. Nothing happened, and for several minutes I sat there motionless and listened. There were definitely noises that came from somewhere down below, things being moved around and perhaps faint conversation. Sometimes there seemed to be louder syllables or words that sounded like "Hey" or "I'm". Just the sort of thing I wanted to hear while alone at sea on a derelict vessel in the middle of the night. What could it be… drug smugglers coming aboard or ghosts? I remembered that Jeff had told me that he too had heard voices out here. He theorized the sounds probably came from doors and various loose items moving with the roll of the ship. The steel hull carried the sounds easily like a sounding board and the wind that moved through the empty window frames produced creepy noises. But what about the footsteps on the stairway? That needed to be investigated.

Armed with an old dive knife and a flashlight, I cracked the bridge door and looked outside. I saw nothing but blackness, a cold stiff wind and the sound of breaking waves. With no visible moon or stars there was little for me to see. My step produced an audible crunch. With the flashlight on I could see scattered chunks of rust that hadn't previously been there. There were more bits of rust on the treads of the stairs leading downward and at the landing was a large rusty piece of steel that had come to rest. Apparently it had broken off the top of the bridge and had tumbled down the stairs producing the noise that had awakened me. That mystery solved, it was time for me to explore further down below.

As I walked down on the main deck past the galley doorway, I heard a loud rustling noise coming from within. I pointed the flashlight into the void in the direction of the disturbance and illuminated the large garbage can at the back wall. Something seemed to move inside it. Cautiously I approached and pointed the beam into the can.

The contents of the can churned excitedly. There were cockroaches and palmetto bugs by the many thousands working the offal. I hadn't seen a single insect my whole time on board and now in this can were enough to supply all of south Florida. They took no notice of my presence and continued boldly with their activities. Bugs normally didn't bother me but the sight of so many certainly surprised me. The container actually rocked with their movements. I turned out the light, backed out the doorway and headed back to the bridge. Maybe it was best for me not to know what went on out here at night.

Five days later the wind finally dropped and the seas relaxed. The overdue *Jefferson* finally arrived, bringing Maurice and my passage back to the real world. When we made the crew exchange this time, I got my belongings off the *Arbutus* and into the inflatible even faster than Maurice had done two weeks before. A hot shower and the fresh food aboard the *Jefferson* were glorious.

We spent another week digging the *Atocha* site with little real effect and then headed back to Key West. Back on shore, it was good to feel an unmoving surface under my feet, and after my stay on the *Arbutus* there was a little swagger to my step. I felt I could take on anything.

Winter of Discontent

December brought more than just northern cold fronts. A competitor for the *Atocha* had arrived and was exploring areas near where Salvors had found artifacts. The principals were Olin Frick and John Gasque. Their project was well funded and they came equipped with vessels and search gear we could only dream of. Their primary vessel, the *Juniper,* was the twin to our own *Arbutus* with one major exception - it was fully functional and in perfect condition. They also had another smaller vessel named the *Seeker.*

The *Juniper* and *Seeker* were using sophisticated side scan sonar to examine the ocean floor just outside of Salvors' legal territory. Mel and Deo went into a panic. While Salvors held a Federal Admiralty law claim to the *Atocha's* remains, the claim was restricted to a specific location, much like a mining claim.

Our great fear was that if Olin and his crew located the *Atocha's* missing artifacts outside Salvors' Admiralty protected area, they might be able to make claim on it. It was possible their superior equipment could find things ours couldn't. Mel's view was that they were nothing more than pirates who were trying to steal the *Atocha's* missing remains from under him. Despite his anxiety, he still kept his sense of humor, calling his competitor's vessels the *Jumper* and the *Sneaker*. Salvors continued in its insolvency, but somehow the money was found to send the *Swordfish* and *Virgalona* out to protect our interests and observe our enemy's operations.

We arrived at the wreck site on December 7th, finding the *Juniper* at work. Ted maneuvered the *Swordfish* near the *Juniper* as it continued to survey the deeper water south east of Salvors' recovery area. Eventually it slowed, going dead in the water, perhaps to adjust or recover its survey gear. Ted pulled within 100 feet of the *Juniper*, taking stock of their activities. I happened to be on the bow with a camera and took pictures of the immaculate ship. Some of *Juniper's* crew on deck laughed and waved at us. After a minute or so of simply staring at each other, Ted turned the *Swordfish* away from the *Juniper* and pushed the throttles forward. Just as we accelerated away we heard five loud popping noises, audible even over the *Swordfish's* bullhorn exhaust.

Olin Frick's *Juniper* docked in Key West

That was a mistake on their part. Ted dutifully notified the Coast Guard and Salvors' office via radio that there had been gunshots fired at us. This action gave Mel's attorneys a starting point to stop the infringement, but it took time to put the case together. For the next several months Olin's vessels and ours played a seaborne chess game, each trying to checkmate the other in the race to locate *Atocha* artifacts in this new search area. Our own survey patterns became reactionary to where the *Juniper* conducted its search. Tensions inevitably rose, and the *Juniper* tried to run down the *Virgalona*.

In court, the *Juniper* crew denied having fired gunshots at us, stating that they didn't even carry firearms on board. Presented as proof of our claims were the pictures I had taken. They clearly showed a figure holding a rifle silhouetted on *Juniper's* bridge wing. A restraining order was issued on behalf of Salvors against the Olin Frick / John Gasque expedition as a result of their hostile behavior in the *Atocha* operations zone. As far as we divers knew, that was the end of it.

The company may have fended off this latest challenge, but from the crew's perspective there was still a lack of real advancement toward finding the rest of the *Atocha*. Morale sagged. In port, heightened fussing began to eclipse the usual nightly drinking arguments. At work or in the local bars the diver's attitudes towards each other were often hostile and combative. Even worse for me, Jeff had finally left for Air Force flight training. We were both envious of each other, he still wanted to stay and find the *Atocha*. I wanted to fly jets. We had become good friends.

The *Jefferson* had finished its contract with Salvors and finally left Key West for a new tugboat job at the end of 1979. Although she had recovered some artifacts, the total finds were disappointing for the amount of effort and money expended. It had always seemed that the *Jefferson* was working on the fringes of an artifact trail, but missed significant concentrations.

Fred had been offered his dream job as a permanent crewmember of the *Jefferson*. He eagerly left Salvors' employ and made sure that we all knew just how much more money he would be making in his new job. Maurice drifted off as well, and new employees of Salvors were indentured to take his place aboard the *Arbutus*.

Our weekly paychecks dwindled to little or nothing. Since the company had no money, maintaining our presence on the site was incredibly difficult. In addition, every two weeks Mel and Ted would have to scrape together enough cash to reprovision the *Arbutus'* unfortunate caretaker.

In port we mostly patched the perpetually leaking *Galleon* and

Ted kept us busy with some painting or other projects. Even though Mel was absorbed in trying to keep what was left of his little company from washing away in a sea of red ink, he was not ignorant of the financial hardships of his remaining employees. Sometimes he would show up with a cardboard case of canned vegetables or soup for the *Swordfish* crew. Cases containing cans of beets or other unpopular foods would be passed off on the next *Arbutus* run.

Salvors' problems were serious and lingering. Subsequent to the 1975 deaths in the *Northwind* tragedy, neither Mel nor Deo were entirely focused on company matters. There had also been much publicity about the State of Florida and the Federal government challenging Treasure Salvors and each other for the ultimate possession of the *Atocha's* remains. Difficulty in raising investment capitol was the result.

The original diving crews had all drifted away when the company's bad fortunes had become too profound. Others had subsequently been hired and again had left over the lack of money. Ted had purchased much of the *Swordfish's* electronics and operational equipment, not knowing if he would ever be paid back. The situation was painfully grim. The draw of the search for the wreck was powerful, but so was the company's economic frailty.

Sensing my own personal funk over Salvors' distress, Leah Miguel made a point of asking me over to their house regularly for dinner. With little interesting food left aboard the boat and the usual clashes between the crew as the only evening entertainment, it wasn't a hard offer to accept. Besides, an invitation to the Miguel's house was a special perk.

Ted and Leah seemed financially independent of Salvors' financial groundings. They lived in a beautiful custom home built on the water and located on nearby Key Haven. It was a place where well-heeled local businessmen and retired military officer's families lived.

On the ground floor were a four car garage, a "mother-in-law" apartment, and a large, well equipped workshop. The shop was where Ted repaired or modified Salvors' worn out equipment. Upstairs, the living areas were breezy and expansive. It was quite a contrast to the smallish, scruffy houseboat on which Mel and Deo lived on in Keys West's Houseboat Row. Mel's leaky houseboat and the *Galleon* were in similar condition; they were both always close to sinking.

In the past, Jeff and I had been out to Ted and Leah's house for social occasions or work related projects. Though Ted was a bit friendlier on his own turf, his usual blunt personality didn't always promote lengthy stays. Now with Jeff gone, Leah sensed that I wasn't particularly enthralled with my remaining shipmates. She went out of her way to make me feel at home. Always cheery with a quick sense of humor, Leah had a genuine concern for my well being at Treasure Salvors. In my off times I found myself more and more in transit to Key Haven.

Mel and Ted regularly discussed how Salvors should continue. We spent as much time magging as financially possible in an attempt to locate new areas to excavate. All the currently known artifact-producing areas on the site had been dug through numerous times over the years with declining success. The *Jefferson's* poor results coupled with the advent of recent competitors seemed to confirm that the company's future depended on expanding the search.

Thousands of linear miles had already been magged around the area where the few *Atocha* artifacts had been found, but the techniques and aging equipment used were now suspect. Though the Frick/Gasque challenge had been a shock, it had also pushed the company to rethink exploring areas beyond the known and comfortable artifact trail. We needed broaden our search, and do it in a more careful and deliberate way.

Ted began planning the new survey's parameters based on wreck features that had been already found and the research of Dr. Gene Lyon, Salvor's brilliant historian and researcher. Dr. Lyon had already located reams of information about the *Atocha* from the Archive of the Indies in Seville, Spain. This Archive was a repository of documents saved from the Spanish colonial period detailing all commerce, and governmental affairs in the new world. Through a long and patient search, Dr. Lyon had located much information; from the ship's construction details, passenger and cargo manifests as well as contemporary salvage attempts. Ted hoped that some of the information that Gene had found might give more clues as to how and where the Spanish themselves thought the ship had met its end.

Of course, any success that Mel had in raising money from new investors would immediately launch our next trip out to sea. Money that came in always evaporated quickly into the company's insolvency abyss, but fortunately for the *Swordfish* crew our operation usually got first dibs. With each brief windfall we'd quickly continue our magnetometer survey in earnest if only for a few days. There was always the awareness that Salvor's future depended on our finding the missing parts of the *Atocha*.

To most of the crew, magging still was not interesting work. Their main desire was simply to dive on any promising "hits" that might turn up. I noticed that the person on board that seemed to be having the best time of all was Ted. He obviously enjoyed working with the charts and survey data, always trying to analyze the daily results. His demeanor was of absolute confidence as he navigated the *Swordfish* on its course, monitored the fathometer and magnetometer displays, made log entries, and directed the crew's activities all at the same time. The only irritation he seemed to suffer was when Mel called on the radio and tried to alter their previously agreed upon survey program.

Ted was a thorough planner and believed that any success we would have would depend on a methodical approach to the search. Mel would always agree to this idea at first, but had little real patience to see the plan completely executed. A day or so into each trip Mel would call on the radio to find out if we had yet found the 'Mother Lode'. Ted would report that we were continuing our thorough survey work in the area as planned, and that so far we had found nothing significant. Instantly disappointed, Mel would then try to get Ted to abandon the project and use his own favorite technique - "wildcatting". This was Mel's term for randomly picking out a place in the middle of the ocean and setting the boat up to excavate. Usually the places that Mel suggested we go were based solely on one of his hunches or a "feeling" he had while sitting in his office or in the local bar. The locations Mel suggested we "wildcat" were sometimes many miles away from the *Atocha's* known artifact trail, seemingly selected by dartboard.

This type of desultory activity placated Mel but grated Ted to the bone. His practical nature required a systematic search through areas that had the best chance of producing results. In this regard, these two couldn't have had more different philosophies - the ying and yang of the treasure-hunting world. Those who knew them best marveled that they both managed to put up with each other.

During one such trip we finished a long day of magging and anchored for the night just where we had quit. The red and purple sunset reflected on the glassy water resolved into a placid starry night with no moon.

Wally had served dinner and the crew talked and laughed amongst themselves between bites below decks. Ted, as was his custom, did not eat or socialize with the divers at night. There was a small pass-through door between the "A" cabin bulkhead and the wheelhouse. Wally would knock on the door when Ted's food was ready. Ted would open the door, Wally would pass the plate, utensils and drink through and then Ted would slam the door shut with the same warmth of a "no trespassing" sign. Within a half-hour or so he would reopen the door and the empty dishes would be balanced on the small ledge beneath the pass-through. Ted would shout 'Thank you" which was Wally's cue to recover them. Ted would then slam the door shut for the night and that was usually the last we would hear from him until dawn. That night though, I had other plans.

While the other crewmembers were absorbed in scuttlebutt between themselves, I quietly left the forecastle and "A" cabin for the back deck. From there I could see the light on in the wheelhouse illuminating Ted working busily on the chart board. Working my way forward along the catwalk I came to the closed wheelhouse door took a deep breath and knocked.

Ted, who was facing away from the door's window glanced briefly over his shoulder at me and motioned inward. As I opened the door there was the sound of disjointed conversations cycling through the marine radio scanner. "Hi Ted- hope you don't mind the intrusion. There's some things I'd like to talk about if you have the time."

He had already gone back to his plotting work and said without turning in my direction "OK, what's on your mind?"

This felt almost as awkward as my job interview with him months ago. "Well, as you know I've been here for a while now and I really enjoy the work. It's not that I'm complainin' or anything but there is a definite routine for the crew and (gulp) I feel a little bit like… I'm in kind of a rut. I'd really like to learn a lot more about the operation – not just the boat stuff, but the survey gear and mapping work as well. I'm pretty keen on learning everything I can about this business. Now that Fred's gone, we don't have a first mate anymore… unless you've got someone else in mind I'd like to apply for the position."

Ted stopped working at the chart board and sat down next to me. "Who told you Fred was first mate?"

"He did- when I first came onboard. Everybody said he was…"

Ted faced me directly with a look caught between laughter and distain. "Fred was not the first mate. That may be what he told you and everybody else, but that is fiction. We haven't had a first mate on board for some time. Fred had been around a little longer than the rest of you but he was no first mate."

While I tried to digest that tidbit he continued. "The last first mate on the *Swordfish* was hired away by a company called Seaborne Ventures …he's now one of their captains. Craig certainly earned that position. When I hired him he was just out of college with no background in salvage or ships at all. Never been out to sea, in fact he wasn't even a diver. Now **he** knew how to apply himself, he got his diving certification and then went on to learn everything here he could. Always interested in learning something more, doing something more. I gave him instruction and books to read on his off time and he worked like a bull."

"At that time Seaborne Ventures' ship, *James Bay,* was working with us for a summer season much like the *Jefferson* just did. They needed a jump-start to help get them in position and working on the *Atocha* site. I had to spend a lot of time with them, helping to sort things out. I turned the *Swordfish* over to Craig and he ran it the whole time I was gone. Did a first class job of it. Now **he** was a first mate worth having. When he had enough sea time, he sat for his captain's license test and passed it. Seaborne Ventures saw what an asset he was and hired him away from me.... gave him a lot more money than we could afford to, and his own vessel and crew. He'll be leaving with them soon for an expedition down in the islands."

Ted picked up his logbook and idly flipped pages while continuing. "I don't coddle people... most of the divers here are happy to stay just that - divers. I'm not going to waste my time training people who really don't want instruction or have the ambition to be anything more than entry-level employees. Cream rises to the top, if you or anyone else wants more than that you're going to have to show me something first."

I didn't quite understand what quality he wanted me to "show" him. Though we talked for a few minutes more, I left the wheelhouse somewhat puzzled but inspired. By the time I reached the back deck and the entrance to "A" cabin, I had begun the formulation of my plan.

It was Ted's routine to rise just before dawn and inspect the engine room equipment and bilges in preparation for the day's work. The whole process took about thirty to forty minutes to complete and was finished before the rest of the crew had even awakened. The crew's alarm clock at sea had always been the slamming shut of the engine room hatch when Ted finished. No one ever stirred until the usual metallic crash.

The next morning Ted was surprised to find me standing in the open engine room hatch, waiting for him. He led the way below and began to describe in detail his morning engineering ritual. From that day on whenever we were at sea, I would help with the daybreak systems inspections and maintenance. In time, some of the other divers sensed what I was up to and decided to join in as a competition for Ted's favor. My new resolve was to always be on the spot at any time whenever anything was required, always ready to learn. Over the months I outlasted them all.

Eventually one morning Ted did not show up at the engine room. I waited for a few minutes in the quiet morning air and then realized his absence was permission to complete the tasks by myself. This trust was surprising, given Ted's normal protective attitude about the machinery on board. There was another side to this coin. My "promotion" into engineering also meant that I would now be held solely accountable for any equipment failures or problems in the engine room. Fair enough.

As 1979 faded and the new year began, some of the biggest troublemakers amongst the crew left or were fired. Ted continued hiring better-educated, more ambitious replacements as best he could. J.C. Borum, a skilled boatman and diver (and James Caan lookalike) was hired. So was R.D. LeClaire. When problems developed in the *Swordfish's* hull requiring that we haul it out of the water at a local boatyard, new employees Don Durant and Kevin Farr joined in on the repairs. Don had come from a family-owned shipping business and knew commercial vessels and diving operations well. Kevin was not long out of college, but mechanically fluent and powerful in the water.

Occasionally I called my former Wisconsin place of work on their

Leah Miguel poses with me aboard the *Galleon* museum while I show off the piece-of-eight coin hanging from my neck.

toll-free line, and talked to my old friend John Green who still worked there. Naturally I always exaggerated how well I was doing in my new vocation and goaded him into coming down for a visit. Being young and bored with his situation like I had been, it didn't take much persuasion to get him to come down. Hard times in Key West looked a lot better to him than winter in Wisconsin and in short order Ted hired John as well.

Curiously, for the entire period of his employment at Salvors, John Green would always be called by his full name. This was quite unusual, as Key West was always exclusively a first name (or nick-name) town. Almost all the people I would come to know there would be known to me (or anyone else) only by one name. Key West was a place to distance yourself from life's previous entanglements and formalities.

From sun up to sun down, boatyard work was hard and dirty. While the *Swordfish* was in the yard, life aboard was about as hard-bitten as on the *Arbutus*. There was no running water or electricity and the day's activities were guaranteed to leave one covered in a pâté of sweat, rust and grease. Most of the crew quickly found friends with apartments with whom they could stay at night. Not having that option, I lived aboard with the other unlucky ones. With no water or cooking capability while the *Swordfish* was laid up, we would sneak into a nearby marina at night and use their showers to scrape off the day's filth. As kind of a comic tribute to Ted's naval career, we painted the *Swordfish* navy grey with a big white and black block number 13 on the bow. "Thirteen" was the *Swordfish's* code number used in radio communications.

Several positive things happened for me. Though Treasure Salvors' pay scale was dismal, the company did have a policy referred to as "coin credit". For every eleven weeks of employment the company presented a crewmember with a grade one Spanish silver piece-of-eight coin from the *Atocha*. Grade one coins were the finest quality, the coins were in mint or near mint condition. Amongst the divers, wearing one of these coins on a necklace was a badge of stature - a professional treasure hunter! In reality, many Key West citizens considered us dirtbags who worked for a quixotic flim-flam man. Nonetheless, those of us who had the coins wore them proudly with our "Atocha Expedition" T- shirts, as we prowled the streets of Key West during off-hours.

I received my first coin and immediately had it mounted for wearing in the expected fashion. When I earned a second, I sold it, and with Leah's help purchased a new Nikon camera and lenses to replace the salt damaged one my father had given me. I loved photography and spent many hours absorbed in documenting my job and surroundings.

Even better for me was that Ted accepted my request for instruction in cartography. Soon I was at the Miguel's house almost every night we were in port. Leah fixed a delicious meal and during dinner we reviewed the events of the day in Treasure Salvors' myopic world. Ted and I then retired to his shop area downstairs for the evening's lessons.

At first these exercises were in making site maps to different scales and plotting parameters. He would give me an assignment that I would immediately begin, often not finishing until very late at night. Leah would usually have gone to bed by eleven o'clock but Ted stayed up as long as it took for me to complete my task. He then carefully examined my math and the results. Once satisfied as to the accuracy of my work he would allow me to go home. While many of the crew spent their nights in Key West's bars, I often was the last one to return to the boat after an evening at the Miguel's.

In plotting our surveys, I came to understand with increasing clarity the site and the difficulties of our present quest. Ted always tried to reconcile the historic information that Dr. Lyon had found in the Spanish archives with what we had found. He believed deeply in the accuracy of the research, but the 1600's physical descriptions of water depths, bottom features and general area location did not seem to match the area where Salvors' made its finds.

We spent hours discussing the writings of the Spanish accounts of the wreck as well as the salvage attempts made shortly after the sinking. Despite all the years of Treasure Salvors work on the *Atocha* site, there were huge informational gaps between the contemporary Spanish writings and the realities of Salvors' limited artifact recoveries. Strangely, the little trail of material that had been recovered so far contained all the major types of artifacts expected from a complete wreck of this type. Representations of general cargo, ballast, armaments, personal items, ship structure and precious cargo had been found, but all in very small amounts. It was as if the company had found the wreckage of a lifeboat, but could not locate the ocean liner from which it had come.

Added to the challenge were Salvors' self-generated problems. During the company's initial efforts on the site, the position of any excavations or finds were often poorly or inaccurately recorded. Trying to interpret where those artifacts had actually been found relative to each other was now very difficult. Also, those early crews dumped junk, steel chains, broken equipment and garbage over the side as they excavated along the artifact corridor. When we re-surveyed those spots with our improved magnetometers and navigational equipment, we got hundreds of "hits" indicating many pieces of ferrous metal on the bottom, which plotted out exactly like a shipwreck trail. Hours or days of excavation

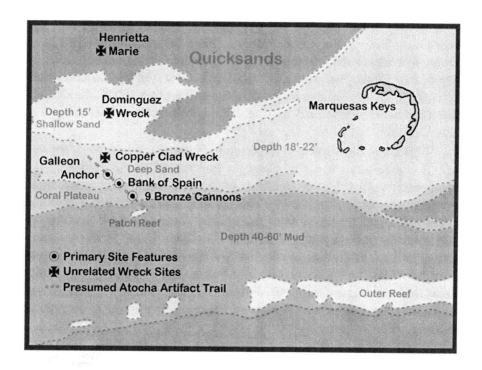

in those spots mostly produced our own 20th century discards. We had become a dog chasing its own tail.

All the *Atocha* artifacts found to this point were along a narrow path that extended about 800 meters along a northwest-southeast axis. At the northern half of the track, water only 20 feet deep covered another 10 feet of loose sand hiding the underlying limestone bedrock. The southern portion was largely a plateau of soft corals and sponges growing on exposed bedrock with a water depth of about 30 feet. At the extreme southeast end of the artifact trail the seafloor made an abrupt plunge down to about 40 feet and progressed gradually deeper. From this point the bottom consisted of a dense mud layer of increasing thickness, overlying more limestone bedrock.

The three major artifact concentrations found so far were evenly spaced along the trail. Their names were rooted in the company's internal language. The first find on the site was a large anchor located during the original magnetometer survey of the area. Although a number of artifacts had been found nearby including the first gold found on the site, the spot was always known as the **Galleon Anchor**.

Several hundred yards to the southeast, just as the deep sand faded into the flat, soft coral meadow was a place that produced the largest number of coins found to date. The area was called the **Bank of Spain**. The coral plateau that continued another 400 yards to the most recent large find was called simply that- the **Coral Plateau**. Nine bronze cannons had been found just where the plateau dropped into the deeper muddy zone. In one of Fate's strangest turns, Mel and Deo's oldest son Dirk and his crew aboard the *Northwind* had found this last artifact feature just days before ithe *Northwind's* accidental sinking. The spot was simply referred to as **The Cannons**.

If a line were drawn connecting these three artifact zones, the line would appear straight. If this straight line was extended further in the southeasterly direction, it would reach another shoal area about three quarters of a nautical mile away. This isolated feature was actually an old silted patch reef that rose from the muddy plane around it to about 17 feet from the surface and was dubbed **The Patch Reef**. As the company struggled back to its feet again after the *Northwind* tragedy, the area of mud between **The Cannons** and **The Patch Reef** had been sporadically excavated and magged as resources allowed. There had been no continued finds, but there hadn't been enough serious work in the area to discount it entirely.

Archaeologist Duncan Mathewson had theorized that the *Atocha* had come in somewhere from this southern, deeper water area. Mel's personal favorite version of that theory was that the ship had come in from the southeast, striking **The Patch Reef** and had sunk somewhere nearby. The artifacts found so far might have separated from the main main wreck, or "Mother Lode", and had been carried away to the location we had found them in.

So far no real results had been achieved in extending the track on either end of the known trail. Further electronic surveys had only relocated Salvors' old refuse, expended bombs dropped in practice during World War II, or other, more modern wrecks. A civil war wreck known only as the *Copper Clad* (from its copper hull sheathing) was found a short distance to the north buried in deep sand. A late 1800's wreck called the *Domingez* had foundered several miles further to the north. A slave ship called the *Henrietta Marie*, a very unique find, was located yet further to the north-northeast. The rest of the *Atocha* "**main pile**", as Mel sometimes called it, was nowhere to be found.

Ted convinced me that Dr. Lyon's archival research held the key as to where we should concentrate our survey efforts. We spent many hours dissecting the volumes of 17th century papers and letters Dr. Lyon had wrested from the vast, mostly uncataloged repository in Seville.

Many of the documents weren't presented as a narrative of the events of the time. They were more a collection of loosely related events, people and activities that surrounded the fate of the *Atocha* and the rest of the fleet that sailed with her. There were tidbits of information about the people and efforts involved in the attempted salvage of the *Atocha* and *Margarita*. I also studied overviews of Spanish history of the time. In absorbing the information, I found unexpected parallels to my own modern world.

Looking Through An Ancient Mirror

At the time of the *Atocha's* sinking, Spain was THE predominate power on the planet. She held interests, possessions and alliances that embraced much of Europe, the New World and the Far East. Bases of operation for trade and well-established colonial settlements for the exploitation of her vast empire covered not only the Caribbean and the Americas but even across the Pacific. Spain enjoyed a cultural zenith and many Spaniards felt that God himself had ordained for them an eternal prosperity. Who could argue this? In addition, Spain was also the grandest proponent of the Roman Catholic Church. The Vatican's intrigal politics and participation in Spanish affairs only gave comfort to Spain's ideas of entitlement.

Ascendancy to this lofty position in world stature was directly due to one event; Spain had simply won the largest lottery in history. The relatively few pesos spent in financing Columbus' speculative voyages of discovery and the subsequent exploration and conquest of almost an entire new continent paid off beyond any bookmaker's wildest dreams. The ultimate grand prize included bulk gold, silver, precious gems and raw materials complete with an indigenous population that could be pressed into forced labor to expedite production.

And there was more. With the Caribbean now a "Spanish lake" and the landmasses of the New World largely under Spanish domain, the new continent was used as a base for trade with the Far East through the Philippines. Gold and silver mined in Mexico was loaded with other European trade goods on huge galleons that sailed across the Pacific from New World west coast ports. The galleons returned from the Philippines loaded with silks, spices, ivory and Eastern exotica, which were transshipped across Mexico and loaded aboard ships bound for Spain across the Atlantic. It was an empire beyond the dreams of Caesar.

To protect her worldwide interests, Spain needed a very large navy and army. Their soldiers and sailors were equipped with the latest in military technology and were trained and competent in traditional European style warfare. Maintaining the huge military put a crippling drain on Spain's already deficit economy, but the world she attempted to dominate was a dangerous place. Many enemies continually challenged her supremacy both in Europe and afar. Ever increasing taxes shouldered primarily by the Spanish middle-class failed to match the government's avalanche spending. Even in that era many of the wealthy or well-connected members of society enjoyed convenient exceptions to their tax obligations.

The Spanish crown's zealous attempts to maintain its power in Europe were often squandered on expensive fiascoes. Spain began a lengthy war with the protestant breakaway Dutch provinces and openly meddled in the internal conflicts in both France and Germany. It attempted an invasion of England in 1588 with the infamous Armada, which completely failed and resulted in loss of equipment and men. A protracted war with France was next. By 1600 the realm had accrued a debt of over 100 million ducats. The one resource that Spain counted on to brace her economy and world power status was the apparently inexhaustible flow of wealth from her New World colonies. And herein lay a vast second front that could be even harder to defend than Her European interests.

The riot of exploration and conquest that began with Columbus had absorbed the lands, cultures and riches of the Americas, to the direct benefit of Spain. With colonial outposts quickly established across the new territories, mechanisms evolved to rapidly shunt this newfound wealth into the Spanish economy. It was critical life support for the country's teetering financial deficit.

The very size of Spain's new world holdings gave her many challengers plenty of room to maneuver. They ranged in scale from fully funded and equipped military operations supported by hostile governments to the small, opportunistic ventures of genuine pirates. The remoteness of Spanish posses-

A model of the *Atocha*

sions strewn across the Atlantic and Caribbean assured that her enemies could traverse the area with little risk of discovery. Their pop-up attacks on both shipping and colonial cities were as random, demoralizing and costly as any terrorist activities in the 21st century. The powerful Castilian army and navy were often poorly suited to challenge quick hit-and-run tactics.

The most vulnerable and tempting targets were the trade ships that delivered European goods to the colonies and returned to Spain loaded with bullion and other New World treasures. By the mid 16th century the crown had decreed that treasure ships would sail in an escorted convoy for mutual protection. Yearly, a great fleet sailed from Spain, driven by the trade winds towards the southern Caribbean along an established route that became known as the Carrera de Indies. Upon arrival the ships would split into two groups; the New Spain Fleet and the Tierra Firma Fleet. The New Spain fleet sailed to Mexico where it collected local gold and silver as well as the silks, spices and porcelain brought across the Pacific aboard the massive Manila galleons. The Terra Firma group sailed to Cartagena and Panama and took aboard the riches of Peru and New Granada.

Ships from the Guard Fleet protected both fleets. Funded and operated by royal interests called the Averia, they were very well armed and provided protection for both merchant's and the king's goods. Paid for by a tax on Indies cargos, this class of ship not only carried cannons and soldiers but also had large holds for the transport of the most valuable types of cargos. Wealthy passengers and governmental officials preferred passage aboard these galleons due to their implied security. The unarmed merchantmen protected by the Guard ships were called naos. Small escort vessels (pataches) also helped to shepherd the awkward procession.

When both fleets finished business at their assigned ports of call, the ships sailed for Havana and regrouped for the trip back to Spain. Havana was the center of Spain's affairs in the New World, and there, final cargos, passengers and provisions were documented and loaded. Hopefully the entire voyage was timely enough that the recombined original fleet made its passage along the Gulf Stream toward Spain before the onset of the feared late summer peak hurricane season. With virtually no ability to predict the onset of a hurricane, the best plan for them was to be out of the area before storms formed.

The *Nuestra Senora de Atocha* was a Guard galleon built in a shipyard in Havana, Cuba. It was actually one of five ships built on a 1616-government contract with shipbuilder Captain Alonzo Ferrera. The contract decreed the specifications and dimensions as well as from where the equipment would come. Though the first three ships were delivered in 1619, in the timeless style of all government contracts, the next two went overtime and over budget.

The *Atocha* wasn't delivered until 1620 and immediately developed serious problems. The main mast shattered on the first attempt to sail to Spain. This caused more delays until a new mast could be fitted. Had the ship been too hastily completed? Already Fleet officials accused Ferrera of using inferior materials and workmanship. The *Atocha* did finally make it to Spain on its second attempt, but required repairs on arrival for serious leaks that had developed in the bow. It was finally ready for the 1622 Tierra Firma fleet sailing and was designated the "Almiranta". This meant that the *Atocha* would sail at the end of the convoy, guarding the rear of the procession.

The *Santa Margarita* also sailed with the Tierra Firma fleet in a position of importance. She was a private venture ship built in Viscaya of sturdy oak, and was destined for the Indies trade. In 1621, *Margarita* had been purchased from her Cadiz owners by the Averia for reinforcement of the treasure fleet. Her own powerful armament supplemented the other Guard galleons on the trip. On March 23, 1622 the Tierra Firma fleet, commanded by the Marquis de Cadereita, departed Spain for the New World. It would be a trip of destiny for both the *Atocha* and the *Margarita*.

Plans for the convoy's schedule went awry upon arrival at Portobello in Panama. On July 1st it was discovered in a meeting of the Marquis, the Terra Firma Admiral Juan Moran, the governor of Portobello and others that neither the crown's treasure nor the merchant's goods had arrived yet for loading. The wait for the tardy overland delivery completely disrupted the important timetable of the fleet. A courier was dispatched with the message to deliver the goods at once. Panama was not the source of the expected bullion; it was just a logistically convenient gateway for the great mines and mints of the viceroyalty of Peru as well as the normal commerce in the region.

In the same meeting, fleet officials learned that the Dutch had established a base in the region and that a large number of Dutch ships had been seen near Cartegena. This news was as ominous as today learning of an active, aggressive terrorist cell in a large American city. The Marquis reacted to the new threat by seizing for crown service a private galleon (moored in Portobello harbor) to act as an additional armed silver galleon. A very experienced and respected captain named Gaspar de Vargas owned this ship, the *Nuestra Senora de Rosario*. It was now the second ship with that name in the fleet.

By July 22 the fleet finished the delayed loading and departed disease ridden Portobello, Panama for Cartegena. Since the Marquis felt the pressure of his flagging schedule, the ships finished their business quickly. In eight days they offloaded the European shipments designated for this port and reloaded with more precious cargo and passengers, ready to sail for Havana.

Calm winds enroute to Havana frustrated plans even further. The fleet didn't arrive until August 22nd. They found that the New Spain fleet had already sailed back to Spain to avoid the storm season. The valuable portion of their cargos had been left in Havana for the Marquis and his Guard ships of the Tierra Firma fleet to transport. The Marquis demanded that the Cuban officials and fleet silver masters hasten the loading of the Havana consignments.

An important meeting was held between the Marquis, fleet officials and the governor of Havana. A decision had to be made about the fleet's return trip to Spain. On one hand, since prime hurricane season was now upon them, wouldn't it be prudent to wait until the dangerous time of the year had passed before trying to return? If the fleet sailed later in the year it just might run into violent winter storms enroute. Perhaps waiting in Havana until next year and enduring the high costs of sitting tight might also be a possibility. There was one great truth that all the participants well understood; Spain was in desperate need of the treasure to keep itself solvent. Careers and reputations could be ruined by the failed arrival of the bullion and freshly minted coins.

There was really only one choice. The Marquis asked the opinion of the chief pilot of the Guard fleet regarding the weather. At that time in Spanish history, high-ranking fleet officials such as the Marquis and in some cases even ship "captains" were not career mariners; they came into their positions through political connections rather than actual experience. It was time to consult with the real professionals.

The chief pilot's response was tilted by the imperfect understanding of atmospheric science of the day. In his opinion, the best weather would likely occur around an upcoming "conjunction" on September 5th. That was a time when the sun, moon and planets would be closest to each other. The new moon occurring prior to this event might produce a period of severe storms. The Marquis decided, based on this weather mojo, to wait until the conjunction.

Final preparations for sailing were made. During High Mass the Bishop of Havana gave his blessing to the Marquis and his officers. That gave some comfort to them but did not eradicate all the tensions over the delayed return to Spain.

On September 4th, 28 ships left Havana at 7 o'clock in the morning for the final leg of the journey home. The day dawned clear and with good winds. The fleet pilots were unanimous in the decision to sail. They were 66 days behind schedule. The Marquis' decision was still tenuous but he ultimately agreed with the pilots. None knew that a monster was already stalking them as it moved through the eastern Caribbean.

As hurricanes go it wasn't that impressive. To slow and clumsy square-rigged ships it was bad enough. The crew noticed the first breath of trouble as they witnessed sunset somewhere between the Florida Keys and Cuba. All night the wind increased and by next morning was at gale force. The ships reduced sail during the day. In the deteriorating weather the convoy began to fall apart. A small ship, the *Buen Jesus*, fell away and disappeared. That evening the wind and waves belied the ability of the helmsmen to steer any course except the one nature dictated.

It was a terrifying and sleepless night for everyone. The normal rhythm of the waves had been replaced by brutal crashing mountains of demonic liquid. Cargos shifted, pottery was smashed and seawater seemed to be coming in from everywhere. Priests gave what comfort they could muster to the passengers over the noise of the wind's determined efforts to remove the masts and rigging. The wind and waves drove the fleet northward. The majority of the fleet was damaged but afloat, passing west of the Dry Tortugas and into the Gulf of Mexico. Captain Vargas' *Rosario, Atocha, Margarita*, a Portuguese slaver and two small escort vessels were blown toward the Florida Keys with their attendant reefs.

The morning of September 6th found the *Rosario*, the slaver and a patache grounded on shallows in the Dry Tortugas. The *Margarita*, about forty miles east, had lost both rudder and foresail and was driven northward. Her crew managed to stop her progress with an anchor not far from the reef line but the line parted in the early dawn. In a seeming moment of divine intervention, the ship was lifted by a huge wave and was carried over the reef undamaged. Her good fortune didn't last. She grounded in the shallow sands several miles to the north of the reef and by 10 o'clock in the morning was beaten to pieces by the ceaseless rollers.

About 7 o'clock that morning, before the *Margarita's* destruction, her crew and passengers witnessed the final moments of the *Atocha* about three miles away. With her main mast and small mizzenmast still intact, the *Atocha* too was lifted by a large wave as she approached the reef. She, unlike the *Margarita*, was driven by the waves directly onto the ragged coral. On impact the main mast collapsed over the side. Waves pushed the sinking vessel past the reef northward until she flooded and disappeared.

Despite the violence of the morning's weather, the storm quickly passed through the area. By the afternoon the wind and sea moderated. A small merchant vessel that had survived the storm found a scattering of 68 *Margarita* survivors clinging to floating debris. Some distance away, the mizzenmast of the sunken *Atocha* still protruded above the waves. Only five survivors clung to it. After they too had been rescued, the little ship made its way to Havana.

The rest of the fleet also limped back to Havana for repairs and to regroup. The Marquis quickly called a meeting to decide what should be done next. The 55-year old Gaspar de Vargas, veteran of The Armada and numerous voyages as chief pilot of the Guard Fleet was of the opinion that the surviving fleet should sail for Spain immediately. The other pilots disagreed, noting the poor condition of most of the ships and the approaching winter storms. General Juan de Lara Moran thought that the search for the sunken galleons should begin immediately. This idea was unanimously agreed upon. Everyone at the meeting knew the critical financial distress the loss of the three treasure galleons would have on the Spanish economy. It would be similar to the stock market collapsing.

The Marquis made the obvious choice. He commissioned the professional Vargas to find and salvage the lost ships. On Sept. 16th Vargas left with five vessels to search for the wreck sites. With help from another Spaniard who had seen the *Atocha's* protruding mizzenmast days earlier, Vargas found the sunken *Almiranta*. Using a sounding line around the wreck he determined the depth of the water, to be about 55 feet deep. His divers couldn't gain access to the interior of the ship as her hatches were still secured from the inside. After the divers removed two small swivel guns from the upper deck, Vargas decided to return after getting explosives from Havana. He prudently left a buoy nearby to mark the spot in case the sunken ship rolled over, and then went to search for the *Margarita*.

Despite his search efforts, the *Margarita* eluded him. Frustrated, he sailed west and discovered the survivors from three ships, including his own *Rosario* on the Dry Tortugas. The grounded wrecks were beyond refloating. To make it easier to recover bullion cargo and artillery he quickly burned *Rosario* to the waterline.

The *Rosario* salvage was nearly complete when on Oct. 5, another hurricane, much more powerful than the first, arrived. Vargas and the survivors took to what little high ground was available to them. Lucky to survive this major storm, they made for Havana at the first opportunity after the storm passed.

With his ship once again re-equipped, Vargas attended to the unfinished business of salvaging the *Atocha*. He quickly established a camp on the islands nearest to the wreck.

Unfortunately, *Atocha's* mizzenmast was no longer visible and Vargas' buoy was gone. The last hurricane had left its mark. His divers held their breaths and groped through the deep, murky, jade colored water. Grapnel anchors were dragged back and forth. Nothing was found. It was exhausting work made worse by the many hours rowing and sailing to and from the camp every day.

Pearl divers from the Margarita Islands were brought in to help. In February the impatient Marquis himself arrived at the salvage camp to try to speed things up. The crew, in an effort to placate him, named the islands Cayos del Marquis in his honor. There was some good news when the divers located two silver bars and other items from the wreck. The Marquis' confidence in the venture returned. He left for Havana, sent more men and equipment to the salvage camp and then returned to Spain with the next fleet.

The following August Vargas returned from the Keys. Nothing more had been recovered. His divers reported ever deepening sand that was covering the wreck was preventing them from finding anything more. Although 100,000 pesos had been spent to recover the treasure, Vargas conceded defeat and returned to Spain empty handed.

Because of our modern day quest for the wreck, I personally understood Vargas' experience. As I read his accounts and felt his same frustrations firsthand I felt a bond with him that bridged the centuries. There was also another person from that time whom I would come to know even better: Francisco Nunez Melian.

Green Flash

Early in the New Year of 1980 the activity level in the company went up another notch with the arrival of the salvage vessel *Castilian* and its owner/captain Bobby Jordan. Bobby was a veteran treasure hunter who looked the part; with his coarse gray crew cut he was as wiry and weathered as his aluminum crew boat. Like Mel, he had cut his teeth on the shallow water *1715 Plate Fleet* wreck sites on Florida's east coast shoreline. Bobby and Mel had recently come to an arrangement for his services as a contractor with Salvors. Treasure Salvors would pay him a per diem salary as well as his expenses, he would be additionally rewarded with a percentage of what he found. Surprising to us Salvors' employees, he wasn't there to find the *Atocha*.

The company's plan still was for our vessels to continue surveying and hopefully locate more of the *Atocha*. Bobby focused his efforts on an entirely new venture, the search for the *Santa Margarita*. We all knew from Dr. Lyon's research that the *Margarita* had sunk somewhere near the *Atocha* in the same storm. He had also located salvage accounts of it from that time. If the *Atocha* continued to prove elusive, perhaps we could hedge our bets by trying for the *Margarita* as well. The only problem was that there was confusion as to where the *Margarita* was relative to the *Atocha*.

In the *Margarita's* survivor's accounts that Dr. Lyon had located, there was a general consensus that the ship had grounded and broken up in water about twenty feet deep. The survivors also agreed that they had seen the *Atocha* sink about one league (three miles) away in deeper water. From here, things got messy.

Margarita's commander had stated that the *Atocha's* direction from his own ship at the time of her sinking was to the east. Based on other information, we of the 20th century questioned the accuracy of this direction. To date, Treasure Salvors' actual *Atocha* discoveries confused things even more. Everything found so far was in shallow water instead of the deep water the Spanish described. Did these artifacts represent where the *Atocha* had actually sunk when the *Margarita* survivors witnessed it? Vargas' divers reported diving in deep water and claimed to have been defeated when deep sands covered the *Atocha's* wreckage. Indeed, Salvors had found deep sand over the artifacts found between the **Bank of Spain** and the **Galleon Anchor**, but this was a shallow area. All of the deep water that Salvors had surveyed so far had a dense mud bottom, not sand. Since we hadn't yet found the majority of the *Atocha's* remains, perhaps what had been found was really a false lead.

Bobby's contract with Mel covered all contingencies. If he located either the *Margarita* or the *Atocha* during his work with us, he would get a numerically small percentage of the total finds. However, based on our knowledge of what was manifested aboard each ship, his actual monetary return could be huge. If he located a completely new wreck separate form the *Atocha* or *Margarita* the percentage of his reward would be an impressive fifty percent.

Castilian was a converted oilrig crew boat similar in appearance to, but one third smaller than the *Swordfish*. It had been modified with the prerequisite propwash "mailbox" system like our boats, and also had a large tower incorporating a diving air compressor built above the back deck. Instead of us-

Swordfish and *Castilian* excavating in close proximity

ing scuba tanks like we did, the *Castilian's* crew used compressed air pumped into long hoses that supplied her diver's breathing regulators. This type of system, called surface supply, eliminates the need to constantly refill and change scuba tanks.

Probably the most novel feature about the *Castilian* was its unpainted condition. The aluminum boat was completely void of paint except for the normal anti-fouling type used to ward off sea growth on its bottom. On the steel *Swordfish*, the crew expended countless hours chipping rust and repainting. The ancient *Virgalona*, with its wooden construction, was almost as bad. Although the oxidized bare aluminum of the *Castilian* wasn't as aesthetic as the *Swordfish's* freshly painted appearance, its simplicity of maintenance impressed us.

The *Castilian* arrived not yet quite ready for its quest. Set up originally for shallow, close coastal work, there had to be modifications made to it in order for it to work further offshore. During this period Bobby was besieged by curious Salvors divers wanting to examine his boat and listen to his stories about the other treasure wreck sites he had worked. He was always courteous and friendly to us although his amicability wasn't universal to all Salvors' employees.

Bobby took an almost immediate dislike of Ted. Perhaps the root of his attitude was in the type of site work common on the *1715 Fleet* sites. On these sites, each of the salvage boats operated as independent entities, even on

Virgalona working in the **Quicksands**.

the same wreck. There was a basic mistrust of anyone else's operations – other boats were often seen as cutthroat competitors for the same finds. Information was never willingly shared and areas where artifacts were located were only divulged for the State of Florida archives as required by the salvage permit leaseholder. Since the sites were literally in the wave wash along the shore, sites were located by lining up shore landmarks such as trees, poles and buildings. Notes on these landmarks weren't shared with other salvors.

Ted wanted to equip the *Castilian* with a LORAN–C receiver for navigation control, the same format as our own vessels. He also wanted to install a VHF radio so we could communicate with Bobby while he was out to sea. Ted expected regular information on Bobby's search areas and survey data reports. Ted's requests and usually officious, uncompromising personality really seemed to twist Bobby. "I am not takin' orders from no damn Navy captain", he was sometimes heard to mutter.

Somehow Mel brought in a little more money. At all hours on the phone, in local bars or talking with visitors to the *Galleon*, he was a tireless and deceptively smooth promoter. We employees got to witness one of his techniques quite often when the salvage boats were tied up next to the *Galleon*. Near the end of the museum tours, he usually appeared with a heavy gold chain around his neck. After intuitively scoping out the most receptive looking individual from the group he would take the chain from around his neck and ask them to hold out their hands. He would tell the story of how the chain had just been found on the *Atocha* and that there were tons of treasure just like it still out there. Wouldn't they like to be a millionaire also when the "main pile" was found? At the same time he slowly dribbled the long, heavy chain into their open palms. His story concluded just as the full weight of the chain dropped into their hands.

That was always the moment of truth. There must have been a certain look, reaction or excreted pheromone that cued him to escort the likely investor back into his office for more discussion and hopefully a signed contract. We could always tell if Mel had had success, for the fresh new investor reemerged on the *Galleon's* main deck and proudly surveyed the salvage boats and divers as the instruments of his newly funded vicarious adventure.

Mel's serious hunting ground for investors was local bars frequented by visiting tourists. His favorite spot was the Chart Room at the Pier House, but over the years also included Two Friends Bar, Rick's and ultimately The Schooner Wharf Bar. Visitors from out of state were defenseless against the naughty, exotic flavor of Key West and a friendly rum and coke drinking treasure hunter talking of National Geographic and millions of dollars in gold and silver. Subsequent rounds of drinks just lubricated the process.

In the long run, sometimes this bar room commerce didn't always turn out so well for Salvors. Mel conceived one famous investment deal literally written on a cocktail napkin when he was desperately short of money. The investor gave Mel $5,000 for 5% of the *Atocha's* finds **in perpetuity.** That meant that every year that Salvors worked the site, 5% was always set aside for that one person. Not bad returns on a wreck thought to have $400 million worth of cargo and artifacts on board.

Always more numerous than investors was the parade of new employees. Most of them had careers with Salvors having the permanence of a tickertape reading. Disgruntled by the lack of money, *Arbutus* duty, or the realities of empty-handed searching, they quickly moved on. A reasonably stable core group of divers did develop; they became the ones who usually filled out the crew roster. Although most had preferences as to which boat they wanted to work aboard, there were no permanent assignments. Kane or Ted selected anyone they needed when one of the boats was to go out.

Virgalona and *Swordfish* continued with expeditions to the *Atocha* site. With still humble finances, the *Swordfish* undertook the majority of the trips. It simply had more capabilities than the slow and shopworn *Virgalona*. Kane went out as money allowed. Our seaborne operations remained magnetometer surveying and diving on likely hits. There were occasions when we experimented with other devices per Mel's request. Sometimes there were homegrown contraptions made of old Radio Shack and Heath kit components, promoted by some hopeful huckster who had convinced the ever-optimistic Mel of the soundness of their science. Magging was bad enough, but there was really some grumbling aboard when we had to take the latest dreamer and his cob invention out to the site.

Inevitably, whenever there was some new piece of gear to test, Mel sent his boats to re-explore his favorite spot; the muddy area between the **Cannons** and the **Patch Reef**. We surveyed this broad area repeatedly with every type of equipment available to us with always the same results…lots of modern junk and bombs.

Bobby finally started his own search. He had brought Craig Boyd to help him and used a different kind of magnetometer from ours called a gradiometer. They had definite ideas where they wanted to work and were generally tightlipped about their theories.

The area in which he was most interested was considerably further east than the *Atocha* track, not far from a sunken destroyer named the *Patricia*, a target ship that the military had bombed for years. Still more comfortable with visual cues than the electronic positioning of the LORAN, he referenced the only objects above water; the distant Marquesas Islands and a few metal pilings surrounding the *Patricia*. Ted had difficulty in getting Bobby to even turn on the LORAN to note positions of any hits.

As the search progressed, Salvors decided that it should have some of its own people on board the *Castilian* to act as liaisons and observers. Don Kincaid, one of the original crewmembers on the *Atocha* project, was re-hired to help in this position. Don was known to some of us because of his appearance in the original National Geographic documentary on the *Atocha* as well as the Bob Daley book on the subject. He had also been aboard the *Northwind* when it sank. Don, a skilled photographer, was well spoken with a confident demeanor. He was not unfriendly, but his demeanor around the current Salvors crew hinted that he considered us also-rans compared to his old shipmates. He sometimes referred to them as "the real crew".

Since Bobby searched areas to the east of Salvors' *Atocha* finds, Mel and Ted decided that we should not only continue to look in the deepwater areas to the south but also to the west.

Fay Feild, another of Mel's old associates, came back to lend his expertise to our efforts. Fay was an electronics genius who originally designed the magnetometers that we used. He was also the only person who could keep the now aged machines working with any reliability. Tall and skinny with a pencil-thin moustache, his attire usually included a panama style hat made from woven palm fronds, a large gold coin necklace and the worst smelling cigars. Since Ted was self-sufficient in mag operations, Fay often would accompany

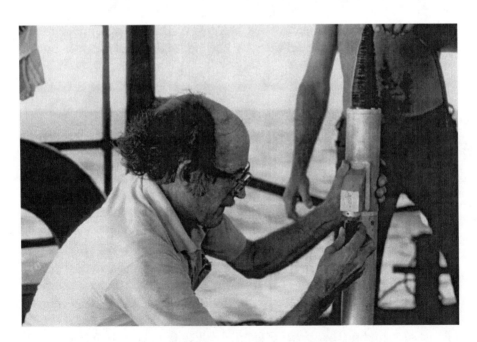

Self-taught electronics guru Fay Feild adjusts one of his magnetometers.

Kane on his trips. At last all the boats were searching. We really started to cover some territory.

Hundreds of "hits" were recorded and dove on by the three boats. All turned out to be modern junk and bombs. Despite the poor results, there was an inertia generated by our efforts. We could all feel it. Paralleling this groundswell was a growing tension between Ted and Bobby. The search had evolved into a competition between mindsets and theories, East vs. West.

Things changed in port as well. Bobby was still friendly to the Salvors divers but there seemed to be problems now between him and Mel. None of us were privy to what was said between them behind the closed door of Mel's office, but Bobby's attitude had changed.

There was an often-repeated company myth, which involved the "green flash". The green flash is a real phenomenon where the last part of the setting sun's disk refracts briefly through the ocean's surface just at the horizon. The conditions have to be just right for it to occur, and it lasts less than a second. The myth was that seeing a green flash meant that you would find gold the next day. None of us had actually seen a green flash. That didn't keep us from melting our retinas every sunset trying to see one. It was something to hang onto.

The tidal flow here was much stronger than Melian had imagined. It had proved difficult to control the movements of his boats when the current was really strong. The divers couldn't swim against it. Harder yet for the oarsmen was towing the submerged diving bell into a good position for searching. The bell did offer the divers some solace from the exertions of fighting the tide and they liked that it could be positioned just above the seafloor for better viewing of the bottom. Other than the problems with the current, the device had worked well and Melian was confident that his investment in it would soon pay off.

He had his salvage boats mark areas with buoys they had already searched. There was little else to reference out here so far from the Cayos del Marquis. The new area in which they were now looking showed promise. There were already signs.

CRESCENDO

Before long, in February, it was my turn again on the *Arbutus*. This time was much more entertaining than the first. I was able to watch the colonization of the *Arbutus* by one of the most unlikely groups imaginable. Inspired by the often-shown National Geographic documentary on the *Atocha*, two quarry men had decided to approach Mel with the idea of installing a large quarry dredge aboard the *Arbutus* to dig in the *Atocha* site's deep sands. The only problem was that neither they nor the crew of young roughnecks they brought with them had ever previously seen the ocean.

Major modifications would have to be made to the *Arbutus* to support the dredge as well as an involved sluice box system they had designed. Many weeks of welding and fabrication would have to be accomplished, all on site. They and everything they needed had to be brought on a barge out to the *Arbutus*.

The quarry men didn't travel light; they had come to stay. They brought tons of equipment and steel as well as their personal belongings. One of the principals even brought with him his physically impaired, plus-sized wife and a large wooly dog. Whatever fantasies of treasure hunting motivated them to undertake this expedition expired upon arrival at the *Arbutus*. The wife and especially the teenage laborers immediately saw their new home as the third circle of hell. Some of the new inhabitants didn't even swim. In a few days they were all tinted with the red ochre of corrosion, grease and sweat that marked the *Arbutus* tribe.

The dredge crew got off to a shaky start. Just days after their arrival one of their major pieces of equipment resigned itself to the waves. A large double pontoon device constructed on shore had been towed out behind the delivery barge and tied behind the *Arbutus* until it was needed. Predictably one night, some passing stormy weather generated moderate seas that caused the pontoon device to break up. Since I was the only good swimmer on hand, I was elected to swim out and tie large hawsers to the individual components so they wouldn't be lost when the structure completely fell apart. I was happy to be taken back aboard the *Swordfish* several days later.

I went back to magging and diving on hits with the other boats. In mid-May the *Swordfish* briefly worked east of the *Atocha* corridor and just south of the *Castilian's* current search area. The area was unusually "clean," but we still registered one strong mag hit in about twenty feet of water. It was interesting enough that Ted decided to dive on it immediately. An inspection dive showed that the seafloor was covered with sand so we put down our mailboxes to excavate. The sand proved to be quite shallow and the source of the mag hit was soon located and brought to the surface.

A heavily encrusted steel or iron object was what had given the reading. Under the encrustation it looked like rows of three-foot long, half-inch square bars banded together. The mass weighed about seventy pounds. We were quite accustomed to finding apparent junk like this, so the item was stored for later disposal.

Not long after we found the "hit", on the 13th of February, the *Castilian* called on the radio. Working a little to the north from our survey zone, they had found something unusual. The *Swordfish* approached her position and several of us boarded our ragged *Whaler* for the short trip to the *Castilian's* side.

John Green was currently aboard the *Castilian* as our Treasure Salvors representative. He and Bobby filled us in on the details. They had been magging with their gradiometer and had gotten a solid hit. They dove the spot and to their surprise found a large copper tub or caldron-shaped object upside-down on the bottom. Since this non-ferrous object couldn't have set off their gradiometer, they searched further. Nearby was a mass of steel or iron bars that were fastened together. They had already recovered the bars, which were now sitting on the *Castilian's* aft deck. The bars were exactly like the ones the *Swordfish* had found.

John led the *Swordfish* crew on a dive to the copper object under the anchored *Castilian*. The water was shallow, barely twenty feet deep. There, in the *Castilian's* shadow was a mottled turquoise-colored shape. It was quite large, perhaps seven feet across and over two feet in height. The sand here was very shallow and in some areas we saw contact between the rim areas and the exposed bedrock. The rim was jagged and torn, suggesting that the object wasn't complete. A line of crude-looking rivets followed the complete circumference near the current rim.

Bobby was convinced it was from the 1622 fleet ships, possibly a cooking pot from the *Margarita*. It certainly looked like the correct vintage but seemed overly large for ship-borne cooking. It could be the remnants of Melian's diving bell. What were the iron bars? There were no good guesses.

We returned to the *Swordfish* and reported to Ted what we had seen, then re-examined our own iron bars. When we carefully removed the concretion we saw there was visible evidence of the nature of their manufacture. The metal wasn't formed perfectly, but had markings and irregularities typical of blacksmithing. It was wrought iron. Old certainly, but how did it relate to that for which we were looking? No one had previously seen anything like this.

The *Castilian* continued to search in the area of the "**cooking pot**". Ted, for his own reasons, decided to continue our efforts back west of the *Atocha* corridor. The *Swordfish's* divers were caught between their belief in Ted's abilities and a growing sense that the *Castilian* might be on the right track. For one brief period it seemed that Ted had won the race for the *Margarita*. As we surveyed a broad area well west of the *Atocha* corridor, we recorded a number of strong hits. The *Virgalona* joined with us and found even more strong readings. The *Swordfish* anchored in the middle of the anomaly field and began digging with her mailboxes. Spikes and fittings from a shipwreck were quickly found. We divers were really stoked and even the usually conservative Ted logged the "*Margarita's*" position.

The elation was short-lived. Closer examination of the artifacts indicated a much later period shipwreck. This was confirmed when further digging uncovered large iron ribs with structure riveted together. Since we had surveyed so much of the area west of the *Atocha* finds with no trace of the *Margarita*, we too would now join the *Castilian* to the east. The competition would continue.

The *Castilian* had been busy. They had found three small anchors scattered to the northwest of the "**cooking pot**". The anchors were about six foot in length, too small for a galleon but of the correct design for the period. They were all oriented in the same direction, as if they had been connected to a vessel riding further to the northwest. The *Virgalona*, searching nearby, found what was thought to be a fourth, but it turned out to be from a later period. We had to separate the ages in the junkyard spread over the ocean floor.

Mel's attorney, David Paul Horan, had been quick to file an Admiralty Law claim on the developing site. It would immediately protect Salvors' interests from interlopers, and give a Federal Admiralty judge authority over the ultimate ownership through the law's framework. It was the salvage law of the sea.

Bobby continued to lead the progression of searching to the northwest. It wasn't lost on us that the track of the anchors and "**cooking pot**" was parallel to the known *Atocha* corridor between the **Cannons** and the **Galleon Anchor**. The three boats continued to find and dive on anomalies, which always seemed to be bombs or junk. Further to the northwest than the company boats, *Castilian* located some possible ballast stone while diving on a mag hit.

Ted brought the *Swordfish* near to the area where Bobby currently worked. The *Castilian* had ceased magging and was excavating. Ted asked me to dive and check out the bottom for any signs of ballast. In short order I was over the side and on the bottom.

The current here was stronger than anywhere else we had worked so far. The bottom topography was different too. The sand depth was irregular, ranging from nonexistent, to rippled miniature dunes. The few sponges and occasional stunted soft corals added visual relief from the otherwise monochromatic sand and bedrock. Here and there were small lumps of dead coral rubble. Bits of loose seaweed zipped past in the aquatic gale.

At first I could see nothing unnatural here. The first few rock-like shapes I picked up were just coral rubble. I drifted with the tide and scanned for something out of place. Then I saw something dark near a piece of rubble. I picked up the barnacle-camouflaged rock. It was a small rounded stone from a riverbed. It was granite and it didn't belong there. Some distance away I found another.

Ted maneuvered the *Swordfish* to pick me up, once aboard I quickly reported to him. Bobby was correct; there was some ballast stone in the area. I expected that Ted would want to begin detailed exploration with the *Castilian*. To the crew's surprise we moved away from the area to other duties.

For the next few trips Ted busied the *Swordfish* with relocating anchors that Salvors had left around the *Atocha* site for later. We repaired broken antennas and pumps on the *Arbutus*, and magged to the south of the *Atocha* corridor. It seemed like we did everything but look for the *Margarita*. Ted seemed immune to suggestions that we might want to join the *Castilian* and *Virgalona* on the new site. We even spent some time digging again on the Civil War era wreck we had found west of the *Atocha*. The anguish felt by the *Swordfish* crew at being left out of the *Margarita* search really peaked one evening.

At night, for protection against wind and waves, the three salvage boats often anchored near each other in the lee of the Marquesas Islands. The crews often visited each other's boats to gossip and talk about the events of the day. Don Kincaid, who now worked aboard the *Castilian* and Bobby, surprised us all when they displayed a number of blackened silver coins they had found while excavating. Most of the coins were in poor condition but some had visible markings comparable with those found on the *Atocha*. They had also found pottery shards and fragments of barrel hoops as well. Was this the *Margarita*? There was nothing definitive yet but the artifacts looked authentic. We returned to the *Swordfish* quite envious and depressed.

As winter faded, spring brought Treasure Salvors more employees. Among them was Australian commercial diver Stuart Preater. Larry Beckman, late of Boeing Aircraft also signed on. Larry brought all sorts of possessions with him, including a houseboat, a homemade submarine and most importantly to Salvors, his own personal side-scanning sonar.

New employees weren't the only thing that April brought. Before heading back to Key West at the end of an uneventful trip, we got a radio call from the *Castilian* to approach them. It was the end of the day and the sun just lingered before it touched the horizon. As the *Swordfish* pulled up near the anchored salvage boat, Bobby and crew displayed three gold bars and big smiles in the reddened light. The attitude of the search had changed.

The new attitude included the deterioration of Bobby's relationship with Mel. We divers could only speculate that there was some contractual dispute going on. In port we sometimes heard a commotion and saw Bobby storm out of Mel's office. Fanning the flames were his continued discoveries made on the site. It seemed the more that was recovered, the worse things got. The *Castilian* found more gold bars, coins and artifacts, but nothing that could positively identify the shipwreck.

The tension wasn't any better at sea. On the eighth of May the *Castilian* was anchored and digging on the new site. We had searched nearby and Ted tried to contact Bobby via the radio to exchange coordinate information. We saw Bobby on the back deck of his boat, waiting for the excavation to finish. Ted pulled close and used the *Swordfish*'s PA system to call Bobby to the radio. Bobby held his hand to his ear and shrugged. Ted pulled very close, using the PA again and gestured at the radio's microphone. Bobby looked uninterested and again indicated he couldn't hear.

Caught up in the effort to try and get Bobby to respond, Ted made a rare mistake. The tidal current had moved the *Swordfish* over one of the *Castilian*'s stern anchor lines. The *Castilian*'s anchor broke loose, she was pushed out of position by the current and rapidly wrapped herself up in her own dragging anchor lines.

Bobby responded by firing a pistol into the air just above us. He screamed curses at the top of his lungs. We withdrew and went back to our own business. The honeymoon was definitely over.

La Margarita

    It was June sixth and the wind had been calm all morning. Standing as high as he could in the boat, Melian could just make out the shape of the diving bell beneath the surface. Even though the water was not so deep, it was too murky to see all the way to the bottom. Earlier the divers had requested that the bell be moved a short distance further to the north. The pieces of wreckage seemed more numerous in that direction.

    A dark shape appeared from under the bell, becoming more defined as it rose to the surface streaming bubbles. The diver waved an arm to catch the well-timed toss of the rope. He was shouting as they pulled him against the current towards the anchored longboat.

    "La Margarita, it is here! Give me some slack on the rope!" Melian put one foot on the gunwale as he and the others aboard the boat watched the diver take a deep breath and disappear again. The sailor holding the end of the line could feel slight tugs above the tides incessant grasp. Suddenly the tugs grew stronger as the diver once again rose to the surface, pulling himself along the rope.

    He hung on the side of the boat breathing hard. "Pull it up-pull it up!" First one, then two men strained to pull the weighted rope. Finally even Melian himself grabbed the line to help pull the object over the side and into the boat.

    The silver bar was blackened by exposure to the sea. Where the bar had been dragged over the gunwale it shone pale. Tears of the Moon. Juan Banon, the slave diver would be granted his freedom for finding the wreck. Now finally after all the self-promotion, political deals, money spent and physical effort, La Margarita was just twenty feet below where Melian stood.

Bobby had run low on food and fuel the previous day and had taken *Castilian* back to Key West. On May 10, the *Virgalona* and *Swordfish* had the new site all to themselves. Kane set up digging in the vicinity where the *Castilian* had been working. He slowly moved southeast, back toward where the anchors and "**cooking pot**" had been found. The *Swordfish* searched with Larry Beckman's side-scan sonar just to the east.

Larry's side-scan sonar mapping the new site.

The sidescan printed out a remarkably clear picture of the seafloor topography several hundred meters on either side of the towed "fish". During the survey we noted three possible areas that looked interesting. We calculated the position of each of the features. Ted asked for a status report on our remaining SCUBA tanks. Several divers went to the dive locker and checked on them. They reported back that all our tanks were near or completely empty.

Ted decided that there simply wasn't enough air left in the tanks for a proper search of the sidescan hits. He knew that divers struggling to do a swimming search against this current would burn up a lot of air. Unfortunately, that meant going back to Key West to refill them as Salvors still owned no working air compressor to fill our SCUBA tanks at sea.

Kane called on the radio and requested another diver. He was short on help, and since *Swordfish* was going back to port to reprovision and fill air tanks, we had spare personnel. Don Durant volunteered, and we transferred him via our *Whaler* and picked up all of *Virgalona's* empty air tanks to be filled.

After the exchange we stopped at the *Arbutus*. Ted wanted to examine a stern compartment for leaks before we left for Key West. John Green was also dropped off because it was his turn to baby-sit the fledgling dredge crew as they labored to get their own operation going.

We were just leaving the *Arbutus* when the *Virgalona* called on the radio. "Unit 13 *Swordfish*, this is unit 7. We need you to bring over your *Whaler*, ours has a problem."

There was lots of laughing in the background during the transmission; enough to suggest something else was up. We headed directly to the *Virgalona*.

Twenty minutes later we pulled to within fifty feet from the old blue boat. Most of her crew, beyond excitement, stood in the bright sunlight on the *Virgalona's* back deck. Kevin Farr held a large gold bar up high in the air so we could see it. Even over the engine noise we heard them shouting at us, and it didn't take long to process what they were saying. There was wooden ship structure and lots of artifacts just under *Virgalona's* stern. The current was so strong now that it carried away the prop blast from her "mailboxes;" they needed *Swordfish's* greater power. The last of their filled SCUBA tanks had just been used up.

Kevin Farr shows off his gold bar.

Ted dropped the bow anchor and backed into the strong current right up to *Virgalona's* stern. He held the boat in place with the throttles until our *Whaler* placed the stern anchors in position. This was normally a risky setup, if any of the two vessel's anchors started to drag; the end result could be really ugly. Fortunately the anchors held and we put our "mailboxes" down, just six feet away from the *Virgalona's*.

Ted shut down the engines so we could talk over the transom to Kane and his crew. They described how they had been working along a trail of artifacts and ballast to the northwest of our present position. Don Durant was one of the divers working in the *Virgalona's* excavation. He had gotten bored and swam off to look around. He swam more than one hundred yards and found some exposed wooden structure on the bottom. On and around the wood were ballast stones, silver bars, copper ingots and lots of artifacts. He surfaced, got their attention and they moved the *Virgalona* to the spot. Although their "mailboxes" were useless in the strong current, they had already recovered several silver bars, a large concreted mass of coins and a gold bar that were lying exposed on the wood. Now they too were low on air.

At news of the find, everyone on the *Swordfish* wanted to jump over the side just to have a look. Ted had other plans. "OK, we're not going to do any more recoveries on or around the structure. I know you're all excited but we don't want to disturb any archaeological information until we can map out the area. Digging or recovery around the wood is off limits for now."

Some of the *Virgalona* divers spoke up. "Ted, there's an area south of the wood where coins are scattered. Maybe if you could just idle your "mailboxes" to gently dust the bottom we could all get a chance to pick up some loose coins with the last of our air. We wouldn't disturb the wood or the main artifact area."

Ted realized he would have a mutiny on his hands unless he granted the request. He agreed to the proposal and with the guidance of a *Virgalona* diver, winched the *Swordfish* into position. In the meantime, all of our crew had raided the ship's dive locker to try and find tanks with any air left in them. A full SCUBA tank held 3000-psi pressure. The most any of ours had left was just under 400 psi. The divers on both boats geared up except for Ted. Just before I jumped over the side with them he grabbed me. "Take the tape measure and get some quick measurements around the wood while everyone is busy looking for coins. Take a look around and get a feel for what's down there."

With everyone under the *Swordfish's* "mailboxes", I swam toward the assumed direction of the wood. With so little air in my tank, I tried to breathe slowly to make it last. It wasn't easy kicking against the current's will. The visibility wasn't good, maybe ten feet at best. For the first couple of minutes all I saw in the shallow sand was scattered ballast. An indistinct dark form to one side turned out to be the wooden structure.

The wood looked like planks from a deck or from the side of a ship lying flat on the seafloor. It was surprisingly clean and quite solid to the touch. There was a large, hefty piece of timber that protruded skyward several feet from the structure. Ballast stones were piled up all around, especially on the western side. Here and there were flat copper ingots that lay on and around the wood with black cookie shaped corroded silver coins everywhere.

I didn't have much air left so I made some quick measurements as I swam around the perimeter of the structure and noted visible ballast, artifacts and their orientation relative to the wood. My regulator started to draw hard and with every breath my tank made a ringing sound, a phenomenon, which indicated it was nearly empty. I swam with the current back toward the sound of the *Swordfish's* engines and in seconds was carried right to where the other divers worked.

The seafloor bedrock where they had gathered was exposed and perfectly flat except for one very large pothole shaped depression. The *Swordfish's* "mailboxes" sent down an invisible column of thrust that gently wafted away layer by layer of sand out of the pothole. In a silvery aura of swirled bubbles, the divers all lay on the bottom around the outer edge of the hole and picked out coins as they became uncovered. I paused just long enough to pick up a small clump of blackened coins that had just been exposed and slowly swam upward while I exhaled.

Wally and Ted both greeted me at the dive ladder. I climbed up, offered my coins and shed dive gear, completely out of breath. Momentarily exhausted, I plopped down on the engine hatch while Wally and I examined the coins I had just brought up. The unfolding events felt like we were watching a movie. The weight and rough feel of the silver "pieces-of-eight" was real though, and I was the first person to touch them in over 350 years. Amazing!

After a minute, I prepared a sketch of what I had seen for Ted. He asked questions while I hurriedly looked for another tank with any air in it. I found one with almost 200 pounds left and strapped it to my backpack. Some of the other divers started to come up the ladder with their own empty tanks. I was over the side in a second and back to the pothole.

There was a gap in the group of divers circling the hole, so I edged into it. Our exhaled bubbles were being forced downward and sideways by the "mailbox" stream, defying normal physics. Between the current and the force of the blowers it was a challenge to hold position, but I was determined to hang on. Since an empty SCUBA tank weighs less than a full one; I should have worn more lead weight. Everyone hand fanned the sand-filled depression as the prop thrust carried away what they had disturbed.

In front of me appeared one silver coin and then another. I greedily

My first coins recovered on the *Margarita* site on May 10th, 1980   Photo: Ted Miguel

picked them up. It was hard for me to hold on to the irregular shapes in this underwater tempest so I started to jam them into my wetsuit. After so many months of hard work to little effect it was an unstoppable reflex.

Our recoveries terminated with the complete exhaustion of diving air and the darkening evening sky. Another back deck meeting was held between the two boats. Ted and Kane agreed that the *Virgalona* should remain on the site to guard the area. For all we knew the entire shipwreck could be right under us. Even though we were over thirty miles from Key West, the numerous fishing and other vessels that regularly traversed the area often observed our activities here. We had not forgotten the recent threat brought on by Olin Frick and Rick Gasque's challenge. Word would quickly spread about the find. With the water depth being so shallow, the artifacts lying exposed and our numerous marker buoys, it would be an easy spot to find.

Ted made the decision that the much faster *Swordfish* would return to Key West that night. We would notify Mel of the discovery and hurriedly reprovision for both boats as well as have all the dive tanks refilled. If we got an early start, we could be back on the site around noon the following day.

We quickly pulled up anchors, and called the *Arbutus* on the radio to ask if there was anything that they needed us to bring when we came back out. Poor John Green had been marooned aboard the hulk three miles away near the *Atocha* site while the rest of us had been bringing up treasure.

There was much excitement in the wheelhouse on the way back to Key West. As everyone talked about what they had seen, Ted noted more details on his site map sketches. Finding the intact wood was completely unexpected, normally organic material is quickly consumed in tropical seas. The only plausible explanation was that it must have been covered with sand and only recently exposed by the currents. Not far to the east of the structure was an area where the ballast stones looked as if they had been intentionally piled up like a low fence. There were no noticeable artifacts amongst the stones. Was this an area where the early Spanish salvors had been working through the ballast looking for the lost treasure?

The next day we returned to the site after a frenzied morning of preparations. Ted maneuvered the *Swordfish* so as to anchor stern to stern with the *Virgalona*, just like the previous day. Food and freshly filled air tanks were distributed between the boats. Ted explained to everyone that Mel was not comfortable leaving exposed high value items just lying on the bottom in such shallow water. His directive was to recover any remaining obvious large pieces such as ingots that lay in plain sight. Excavation on or around the wood was still off limits. With that understood, it was time to work.

Two more silver bars and a number of copper ingots were found lying exposed on the ballast stones. The silver bars were especially important because of the serial numbers and weights stamped on them. These numbers could be correlated with the *Margarita's* original cargo manifest that Dr. Lyon had found in his research. If the numbers and weights matched, that would be definitive proof that this was indeed the *Santa Margarita*.

Once the exposed ingots had been recovered, the divers from both boats continued the excavation of yesterday's coin pothole. Again, under the stream of the *Swordfish's* mailboxes, we all gathered around facing the ever deepening form of the depression. I picked a spot near Stuart and Larry. I was wearing much more lead weight on my weight belt, which helped me hold my position against the assertive turbulence.

The pothole we faced was actually created long before there were Spanish galleons. Geologically it was called a solution hole, a feature created by acids from decaying plant material dissolving the soft calcareous rock. Though ancient sea creatures had originally created the rock, the ice ages had changed the ocean level dramatically. For a time this seafloor had become dry land crowned with heavy foliage. Tannic acids eventually pooled and razed portions of the rock creating the potholes. Thousands of years later we were diving over the once-again-sunken landscape, recovering artifacts that had fallen into the potholes. Centuries of storms and ocean currents had shifted the lighter items into the nearest solution hole where they stabilized and were covered by sand.

During my first dive on the pothole I found part of a silver plate, hundreds of loose coins and even some large clumps of coins fused together by silver sulfide. Suddenly right next to the edge of the depression was - could it be? Incandescent even in the filtered underwater light - a shiny piece of gold.

For some reason I hadn't expected this. It was a large twisted link from a "money chain". In 17th century Spain there was a law, which declared that any wealth that could be worn as jewelry would not be taxed. Because of the law, the wealthy often wore large gold chains. This not only served to announce their social status, but had a practical side as well. Links of the chains could be pried off, and based on the bullion weight of the link, used as money. The chains were made of very high quality gold (twenty-three and three quarters carat) just like the coins of the day.

The link I found was about an inch long and looked like it had just been made. No concretion or discoloration hinted at its age. It might have once been someone's pocket change. I edged over to show it to Stuart whose wide eyes barely fit behind his mask. Larry joined the huddle with a large gold ring with an emerald mounted in it that he had just found inches away. The ring, like the chain link, looked remarkably new. It was difficult to conceive that these things had lain here over 350 years in only eighteen feet of water.

We all spent the rest of the day underwater around the pothole, surfacing only to change tanks. The shallow water depth allowed us virtually unlimited bottom time. As the wind and seas picked up at sunset, both boats elected to spend the night in the protective lee of the Marquesas Islands. Everyone was exhausted after the day's cocktail of adrenaline and prolonged physical effort.

The wind stayed strong throughout the night. The next morning brought rough seas that reduced underwater visibility. We tenuously anchored again over the pothole but the recoveries from it began to thin out. Mel was ferried out from Key West aboard a fast boat so he could briefly inspect the site. Already the area around the structure was being referred to as the **Margarita Main Pile.** The name stuck.

Our remaining air tanks lasted for a few more dives. Mel left as the weather worsened, and we all decided to head back to Key West. The *Swordfish* matched the slower *Virgalona's* pace so both boats would arrive back at the *Galleon* simultaneously.

The word about our big find was definitely out. As we entered the harbor we could see a huge crowd waiting for us on the *Galleon*. The *Swordfish* tied up first, and a human tidal wave crested with still and TV cameras washed aboard. When the *Virgalona* tied up next to us, the crowd jumped aboard her as well, and the old boat began to sink from the sudden weight. Ted and Mel quickly limited those on board to just the media. The rest were herded back aboard the *Galleon* where they watched as the discoveries were displayed for the cameras.

Bobby Jordan appeared from the throng and assumed a place in front of the cameras as if he had made the finds himself. Although this caused much mumbling amongst the divers, Bobby enjoyed the spotlight.

Mel, the eternal showman, set up a scale on the *Galleon's* main deck for the media's benefit. Piece by piece we weighed the recoveries: 850 lbs. of silver, 1305 lbs of copper, coins by the hundreds and of course the gold. We divers were suddenly transformed from Mel's gang of dirtbags to Key West's favorite sons. All of us took turns being interviewed. The subtropical afternoon sun was never so bright.

That evening there was a big party at the Pier House Hotel. Mel even got all the divers rooms there for the night. It was an extreme gesture for a man who usually couldn't make payroll. The celebration lasted for some until the wee hours, but I was just too knackered and soon retired to the precious luxury of my air-conditioned suite.

The following day the company tried to regroup. Rising dissonance between Bobby and Salvors became quite public during an examination of the silver bars' markings and a later crew archeological meeting. Bobby no longer confined his dissatisfaction with his percentage or the publicity about the discoveries to Mel's office. He stormed off the *Galleon* and shouted at the bewildered tourists "Mel Fisher didn't find this wreck- Bobby Jordan did, Captain Bobby Jordan!" There had always been respect for Bobby and his abilities amongst the divers, but his outbursts won him little sympathy. One issue had been definitively settled. The silver bars did match weights and markings from the *Santa Margarita's* manifest. It was the *Margarita* that we had found.

The next few days were a blur. The *Galleon* was as filled with people as it was with harbor water. No longer did Mel have to work the crowd for likely investors; they lined up to throw money at him. Ex-employees and long gone associates reappeared, scratching around for some opportunity to get involved again.

Others with real value also returned. One was Bleth McHaley. She had performed public relations and paralegal duties for Treasure Salvors in its early days. After hearing about our finds she returned, seamlessly assuming her responsibilities as if she had never left. Duncan Mathewson came back as well. He was an archaeologist who had also worked with T.S. in its early days, and had tried to better organize the company's efforts. Duncan had analyzed the finds that had been made on the *Atocha* site, and had theorized that the main section of that wreck was somewhere in the deeper water. The eventual finding of the bronze cannons at the edge of the drop off had given more credence to his hypothesis. He had come to assist on the new wreck.

On our next visit to the *Margarita* site we took out most of the Fisher family, several visiting members of the National Geographic staff and Bleth. They all had brief dives to explore the structural area while the *Castilian* dug nearby. Back in port, the local bank employees got a real thrill when we brought in much of the silver bullion for temporary storage in their vault.

New investment money coming into the company allowed the *Swordfish's* beat-up *Whaler* to be fitted with a brand new outboard motor. A number of new, state-of-the-art metal detectors were issued to the boats as well. Despite all this newfound celebrity and extravagance, we divers still weren't exempted from having to spend hours patching up holes in the damn *Galleon*.

Bobby had another good trip with the *Castilian*, and on his return showed off some more gold bars that had been located near the **Margarita Main Pile**. Perhaps because of his recent outbursts his reception among the Treasure Salvors personnel was unusually cool.

At last the *Swordfish* was ready to head back out to the site. We were especially eager to try out a new hand held "blower" that Ted and I had concocted during our evenings in port. It was constructed out of an electric trolling motor, a bicycle handle bar and a bilge pump float switch. Hopefully it would be useful for blowing the sand out of deeper potholes. Everyone itched to try out the new, sensitive hand-held metal detectors.

We spent a few days on the *Margarita* site and experimented with the equipment. The detectors worked better than imagined and located hundreds of coins and other artifacts in casual exploration beyond the perimeter of the structure. The portable blower was almost too powerful. We found that unless weighted heavily, the diver operating it would be propelled away from the solution hole he tried to excavate. During our experiments, we discovered a familiar object that lay right next to the structure; an intact World War II bomb.

La Margarita

Gold bars, a quartz cannonball and encrusted sliver coins and plates found near the ***Margarita* Main Pile**

Despite the crew's desire to begin serious exploration of the new wreck, Ted decided to take the *Swordfish* back to the deep water near the *Atocha* site. Examination of Larry's sidescan sonar paper readout showed that one of the anomalies we had marked hours prior to Don Durant's discovery was in fact the ***Margarita* Main Pile**. The coordinates we had extrapolated for the anomaly were exactly those of the actual structural area. Had we had full air tanks on board for a proper search of the area, we would have dove it at the time. With this knowledge, Ted was determined to try his luck with the device for the rest of the trip in search of the *Atocha's* "main pile". It was a worthwhile gamble, but unfortunately nothing remarkable turned up.

On the next trip we were back on the ***Margarita* Main Pile**. The one-day mission began by our setting up benchmarks and a grid system around the structure for mapping and recovery purposes. Our reprise archaeologist, Duncan Mathewson and I measured the wood while a visiting photographer from National Geographic documented our activities. With a better understanding of the layout of the exposed timber, we could prepare a plan for detailed mapping. Our study of the construction might determine what portion of the ship the wood represented.

In the midst of perfect weather and with a valuable wreck to explore, the *Swordfish* was once again held up in port by the need for repairs. The heavy frame that supported the canopy over the back deck had rusted so badly that it broke and partially collapsed. In the summer the canopy was as vital to our operations as any piece of equipment on board. The sun's effect on the anchored boat's steel deck and superstructure at this latitude was almost lethal, and the canopy's airy shade offered the only comfortable place to work during daylight. Several other crewmembers and I began to quickly fabricate a new steel framework; we welded it together in the blast furnace environs of the *Galleon's* marl parking lot.

The framework was completed and installed in three hellish days. Just as we finished up, something strange occurred. On May 26th, from the rear of the Pier House Hotel, Don Kincaid had observed the *Castilian* returning to Key West at the end of her latest trip. Instead of coming into the harbor and tying up next to the *Galleon*, she anchored in the channel off the Pier House. After a short time she raised anchor and departed Key West quickly without contacting anyone at Treasure Salvors. On board were Bobby and his crew as well as two T.S. divers, Frank Moody and R.D. LeClaire.

By the following morning an interplay of phone calls, aerial searches, lawyers and local authorities eventually produced the *Castilian* and its crew, miles up the Keys. Frank and R.D. each had stories that were diametrically opposed in detail, but agreed fundamentally on what had occurred.

The wooden structure found on the **Margarita** **Main Pile** was eventually determined to be the side planking from the ship's sterncastle. Photo: Pat Clyne

The *Castilian* had enjoyed a fantasy trip. As the crew worked near the **Margarita Main Pile** they had found numerous gold bars and discs, 2 silver bars, gold and silver coins and other artifacts. Upon returning to Key West they had anchored as observed by Don. Bobby then called his wife on the radio and had a short, almost cryptic conversation. At its conclusion he had taken the *Castilian* directly to his financial backer's house on Summerland Key. There the T.S. employees were shepherded from the boat and into the main house.

A U.S. Federal Marshall had come to the residence as requested by Bobby's people. The marshall's function was to examine the recovered artifacts aboard the *Castilian* and to officially establish an Admiralty Law claim on the wreck from which the items had come. Bobby and his backer tried to represent the items as coming from their own, newly discovered wreck. The marshall was ignorant of the fact that the objects were actually from the *Santa Margarita,* which already had an Admiralty claim filed on it by Treasure Salvors. When the official proceedings concluded and the marshall departed, there was a great deal of celebration that night by the principals who even saw fit to include the bewildered T.S. divers.

Bobby's scheme quickly collapsed. Made aware of the deception, the authorities rescinded the bogus Admiralty claim and confiscated the stolen artifacts. Legal proceedings were begun against Bobby. He would never again return to the *Margarita* site, nor receive anything from it.

From my perspective it was hard to understand how this serious and competent salvor had thrown away his chance even before we had begun to explore the wreck's potential. Now he would get nothing for his efforts but legal trouble. First arguing for a larger share than in his contract, and then against the evidence that the new wreck was the *Margarita*, he had tried to stretch his percentage to fifty percent of the find. Was it a combination of gold fever and bad advice that pushed him to finally claim the entire site for himself?

Gossip about Bobby's misadventure didn't last long in the lower ranks at Salvors. There was already an overload of other stimuli to deal with. The company morphed daily with the influx of money, new equipment and new activities.

Adding to the confusion, the Mariel, Cuba boatlift was going on all around us. Nearly one hundred thousand Cuban refugees and their boats arrived in Key West and the lower Keys. The Coast Guard and INS tried to process them as they arrived, but many escaped, leading a troglodyte existence. The junk piles and wrecked cars on the perimeter of the *Galleon's* parking lot were some of their favorite nesting sites. Confiscated or abandoned boats and small ships from the boatlift choked Truman Annex harbor, were piled on the seawall and sank at the docks.

We were happy to trade the craziness in Key West for more time on the *Margarita*, even if it was for only a day. The *Swordfish* made a quick trip out to finish setting up a grid over the exposed wooden structure, and placed a string of numbered concrete blocks leading from the wood out five hundred meters to the northwest. Both were going to be used as controls for an upcoming photographic survey. Two new/old T.S. employees, Pat Clyne and Tom Ford, made the trip with us.

Both Tom and Pat had been with Treasure Salvors in the early years and our recent success had drawn them back. Pat was fluent in underwater photography and had once found a silver bar on the *Atocha* site. The only thing I knew of Tom was that he had helped recover the drowning victims of the *Northwind* sinking. During the course of the day we talked of his strong interest in the archaeological aspects of the *Margarita*. All of us would be involved in the upcoming photo mapping.

The real highlight of the day's trip occurred that evening. We arrived back at Key West after dark, and as we made our way through the harbor to the *Galleon* we noticed that something was different. From the parking lot there were massed car headlights aimed at the ship. From our point of view the familiar silhouette appeared distorted. The masts were at an extreme angle and when we pulled up close we saw that the *Galleon* leaned over hard against the seawall. Despite the daily patching, the hull had deteriorated so badly that the bilge pumps just couldn't cope. Optimist Mel figured that it would just sink gently onto the bottom of the shallow harbor, allowing continued use of the upper decks and sterncastle offices. Instead of settling upright, the *Galleon* had rolled over, crushing its starboard side in against the seawall. A number of Salvors employees braved the steeply tilted awash main deck to rescue anything of value. In the midst of everything else, our headquarters had sunk. Here was one galleon dive we were happy to never again have to make.

While working on the sunken Santa Margarita, Melian recovered bullion that he declared was not from the Margarita itself, but from a nearby, unidentified wreck he had also found. His Royal Charter granted a certain percentage of the profit for recoveries from the lost 1622 fleet ships and an even greater reward on any unknown shipwreck sites he might find. Rumors and eventually accusations emerged against Melian in Havana, claiming that he had not accounted for everything found. It was also implied that Melian's powerful friend, the Governor of Havana, was in on the deception.

Neither had to stand against formal charges of embezzlement, they were too politically connected for events to advance that far. The issue was more sidestepped than resolved. (Note: Even with all the electronic survey resources available to twentieth century salvors, no older shipwreck was ever located in the vicinity of the *Santa Margarita*.)

The *Galleon* museum the morning after its sinking

Rogue Wave

With the *Galleon* on the harbor bottom, all company personnel were pressed to recover any company records, equipment and display items that hadn't been ruined. This also presented employment possibilities for some of the ex-employees and other hopefuls who anxiously hung around. A guesthouse several blocks away on Simonton Street was rented as our temporary new office.

As with much of life, timing was everything. If we hadn't found the *Margarita* there would have been no money to relocate the company. Salvors normally wouldn't have been able to survive without the meager receipts from tourists who visited the *Galleon* and Mel's efforts to encourage them to invest. Now we were flush.

Both the *Swordfish* and *Virgalona* were soon equipped with new diesel powered air compressors. Now our trips would no longer be limited by lack of filled SCUBA tanks. In order to preserve as much deck space as possible, we mounted the *Swordfish's* new compressor on the cabin roof just behind the back windows of the wheelhouse. It was somewhat protected there from salt spray but added another thick layer of noise to the operation of the boat.

Strange new gossip went through the crews as we cleaned out the *Galleon*. It was reported by some of the *Virgalona* crew that Kane was quitting as captain. This made no sense to any of us, but he wasn't around to ask about it. Interjected in speculation was the rumor that he wanted an old family friend called "Momo" to take over and run the boat.

Momo's real name was Demestones Molinar. He was a Panamanian diesel mechanic who Mel had met years earlier on a treasure hunting expedition in the Caribbean. Momo had "stowed away" aboard Mel's boat when it returned to the U.S. When Mel began to work the *1715 Plate Fleet* sites, Momo was there as well. He learned to dive and was absorbed into the processes of Mel and Deo's efforts on those wrecks, helping recover treasure and artifacts.

When Mel made his move to Key West to look for the *Atocha*, Momo served as captain of the *Virgalona*. After the *Northwind* sank and Treasure Salvors' finances ebbed, he had left to find other work.

Another odd occurrence concerned Don Durant. Since the day when he had first spotted the **Margarita Main Pile**, his behavior had changed. Normally practical and even-tempered, he had become extremely agitated and edgy; constantly pacing like a caged animal. Nothing seemed to calm him. He soon left Key West and Treasure Salvors to rejoin his family in New Orleans.

Toward the end of May, Ted and I drove to Key Largo to pick up an underwater photographic unit called a Ribicoff Pegasus. It was a self-propelled torpedo shaped device that a diver could "fly" underwater like an airplane. It had a bank of strobe lights and cameras mounted on the front of it and would be used to document the structural area on the *Margarita* site. The owner and inventor, Dimitri Ribicoff, was a contemporary of Jacques Cousteau. On a future trip he would join us to operate the machine.

On the way back from Key Largo we stopped to have lunch. In the middle of an unrelated conversation Ted suddenly changed the subject. "Very soon I'm going to be too busy to take the *Swordfish* out any more. I'm offering it to you. If you don't want it, Kim Fisher will come down and run it."

WHAT??!! Did I hear that right? After I thought over the statement to be sure I'd heard correctly, I quietly asked, " Why would you be too busy?"

As he worked a toothpick, he leaned back in his chair. " We're looking for a new boat for Kane to run. There's one up in Charlestown that might fit the bill. I'm going to have to go up there and take a look at it, and if we buy it I'll be bringing it back down here with Kane to make sure everything goes OK. It'll have to be hauled out of the water for refitting and then we've got to fabricate a winch system, mailboxes and all the other stuff. It's going to be a big project and I want to make sure it's done right. I'll be hiring a crew to help in the boatyard and I'm going to need to be there to direct what they do. I can't be driving the *Swordfish* and do that too, so I'm givin' you first option."

I felt gutshot, even at a time when the unexpected seemed to happen almost every day. Kim Fisher was Mel and Deo's oldest surviving son. Like his little brother Kane, he had grown up and worked with his parents on the *1715* sites. He had captained the *Southwind*, the sister ship of the ill-fated *Northwind* during the early years of the *Atocha* search. For his own reasons he had left T.S. and moved his wife and growing family to the Midwest where he studied to be an attorney. I had met Kim briefly during a winter break, when he and his family were visiting Mel and Deo. He seemed to be a nice guy with an easygoing demeanor.

I didn't have a problem with Kim, I barely knew him. On the other hand I knew that with any management change aboard would come a new routine. Though Ted could be frustrating at times, I respected his way of doing things; the boat and its equipment were always squared away and ready to go. Unlike the slapdash preparations that often preceded the *Virgalona's* trips to sea, Ted made sure our maintenance and repairs weren't left to the last minute. We always had all the equipment and supplies we needed for the trip. There was continuity to the way things ran and a professional pride bonded the crew. Kim was more than qualified, but if I stepped up to the plate there might be less of an adjustment for all of us.

Our biggest gripe about Ted was his apparent ambivalence over working the *Margarita* site. To us, it seemed that he was often drawn away to other projects higher on his priority list. If I captained the boat, development of the site would be my first and only mission. The rest of the crew were just as eager. Was I ready though?

I certainly knew both wreck sites as well as anyone from all the survey charting I had done with Ted. He had thoroughly instructed me on the ship's engineering and equipment. On occasion, Ted had allowed crew members to practice maneuvering and docking the *Swordfish*. I had taken my turn at this exercise, but had spent no more time at it than anyone else.

There was another big potential problem. Up to now, I had been equal in status and rank to everyone else who worked on board. While I had a personal goal of achieving first mate status, Ted had never made any indications that he was fast tracking anyone for that position. I would be skipping that intermediate step and would be completely in charge. What problems would this cause?

"Ted, if you think I'm up to it, I'd like to give it a shot. I'll try to do my best."

"You'll do fine. Remember, your authority on this thing comes down from me. Mel and I have discussed the situation and we think you're the best man for the job. For now though, I want you to keep this conversation under your hat - don't tell anyone about it. At the right time I'll inform the crew. Just carry on as always."

"When will this happen?"

"I'm not sure yet, but soon. I'll give you a heads up in plenty of time."

I had plenty to think about. In the back of my mind was an endless dialogue about the crew's future acceptance of my new position… and my ability to perform it. An additional challenge existed in all of the old hands who had showed up expecting their jobs back now that the *Margarita* had been found. There was one particularly bitter ex-employee we nicknamed "Sphincter", who felt entitled to an immediate, elevated position within the operation. When that didn't happen, he entertained himself by making numerous loud, derogatory comments about the current crew. There was little doubt how he would feel about my promotion.

A brief distraction from all this was the process of dismasting the *Galleon*. There was was fear that the heavy masts and rigging would cause the ship to roll over even further. The decision was made to remove them with a crane. I remembered that Mel had once mentioned that when the *Galleon* had been built that he had "stepped" the masts with Spanish coins for good luck in the old sailing ship tradition. That meant that a coin was placed under the foot of the mast when it was put into place. I assumed the story was B.S., but was still intrigued.

When the small mizzenmast was pulled out, I scrambled into the still dry, fun-house confines of the sterncastle where the mast's base had rested. Sure enough, there was an eight reale 1715 silver coin lying on the "step". If a silver eight reale coin had been placed under the least important mast of the ship, what about the main mast? I showed the silver coin to Mel who only laughed. Since the forward mast had been removed several years earlier, the only other possibility of more coins being found was under the main mast. While that mast was being pulled I devised my plan.

Early the next morning I climbed onto the *Galleon's* deeply sloped and awash main deck. Some of the other *Swordfish* crew were aware of what I was doing and helped carry my dive gear. I was going to enter the *Galleon's* hull through the empty mast hole in the main deck and try to swim into the bilge area and locate the mast step. The visibility inside the *Galleon* was near zero, so I brought a powerful flashlight and was tethered to a safety rope.

The deck mast hole was so small that I had to slide through it and then have my tank handed down to me. With a last upward glimpse of my friend's faces framed by the hole, I descended into the darkness. It was toxic Braille diving inside. Pieces of the display cases, dead rats, mannequins and other undecipherable museum junk half-floated in the black void. It took at least fifteen minutes for me to locate and identify the mast step by feel. Even with the flashlight beam and my mask right against the step, I could see little. The best I could tell was that it was rusty steel. No coin was visible. Just before I gave up and made my way back up the rope, I rubbed the corroded surface of the step.

Near the center I saw a yellowish glint. I rubbed again and looked as closely as I could. Sure enough, there was a small gold coin. I pried it out of the rust and gently returned upward along the line. Just before my head cleared the surface at the mast hole, I extended my hand with the coin upward so it would be the first thing my comrade's saw. They yelled and shouted as my head cleared the water.

We passed the coin around for inspection after I was out of the water. It looked as if it was a two-escudo gold coin from the *1715 Fleet* sites. It was stained somewhat by its contact with the rusty steel of the mast step so that evening I took it to Ted and Leah's house for cleaning. Gold is a noble metal, unaffected by acids. Ted put the coin in a mild acid solution to remove the stains while we ate dinner. I felt pretty good about myself, especially enjoying the other diver's envy.

After dinner I went downstairs to Ted's shop to check on the coin. The acid had done its job almost completely; there was only a tiny bit of rust still visible. I held the glass jar up to the light and gave it a shake. The jar turned into a gold tinted version of a glass ball snow scene. The coin had been a gold plated replica. It was a lead doubloon. In my head I could just imagine Mel's nasally laugh.

The *Swordfish* made several more trips to the sites without Ted mentioning anything more about our conversation. While we were on the dock preparing for our next voyage, Ted pulled me aside. "Tomorrow you're officially in charge. Get with me in the morning and I'll give you checks for groceries, fuel and ice and an operations fund for anything else you may need. The weather looks like it'll be good. As soon as you're ready you can head out."

So that was it. I wasn't sure whether to say something to the others about my new position or not. I remembered that Ted had told me that he would inform the crew at the right time. After thinking about it I decided to say nothing for now and see if Ted made good on his statement by the following morning.

I felt somewhat apart now from the rest of the crew and I watched their activities in a different light. Ever since the company had become financially flush, we all worked hard to improve life aboard the *Swordfish*. The shower stall was made functional, a new hot water system was installed, the galley was updated and new cabinets and personal storage areas were fabricated. These projects not only added to our comfort but also had started to unify the diver's strongly divergent personalities. I hoped my promotion to captain didn't fracture this fragile bond.

The next morning, on June 17th, Ted was true to his word. Everyone seemed to know I was the *Swordfish's new* captain. We were too busy getting ready to go to sea to talk much about it but so far everyone on board seemed supportive. While Wally went grocery shopping, I piloted the boat over to the fuel and ice docks to take on what we needed there. With that completed we picked up Wally and the provisions back alongside the sunken *Galleon* and then headed out through the harbor. The *Swordfish's* exhaust echoed proudly off the stone jetty.

The new captain of the Swordfish   Photo: KT Budde-Jones

Once we had cleared the channel, I turned west toward the wreck site. I turned the helm over to one of the crew and sat down to make my very first log entry. At some point everyone came into the wheelhouse to congratulate me on my promotion. Their well wishes got even more enthusiastic when they learned that my only plan was to excavate the *Margarita*. I already sensed that some hoped Ted's stand-up management style had been left on the beach with him.

I reviewed our site maps on the chart table to figure out a plan of action. From what we had already found on the *Margarita*, the direction of the artifact scatter paralleled the known *Atocha* material.

I knew that Bobby and Kane had both worked along a line northwest of the **Margarita Main Pile**. Though we had found a large concentration of artifacts and ballast stone around the "pile," it had become obvious that the whole wreck was not represented in this area. There should be far more ballast stone and artifacts, which included large items such as cannons and other armaments. For this trip I decided to cover all bets by exploring further to the west, south and east of the structural area with the *Swordfish*, while sending a team in the *Whaler* to the north.

After two and half-hours of travel, we saw the small marker buoys scattered around the *Margarita* site. I was quietly relieved that my navigation was good enough to find this remote spot. I pulled the throttles back and let the tide continue to carry us toward the main pile area. With the noise level now low enough that I didn't have to shout, I briefed the crew.

"OK, we've all been anxious to get out here and really work this site… now's our chance. My plan is to set up the boat so we can explore a little west, south and east from the wood structure area. We'll be using the blowers to "dust" the bottom as we go. I'm also going to have a detector team free range just to the north of the wood as well. At the same time Wally will take the *Whaler,* and tow a diver just above the bottom further to the east and north. I want him to mark any visible ballast or artifacts with buoys… let's find the visible outer perimeters of the site. All of us are going to be very busy, there's going to be no time for sitting around. I'm sure most of you have felt like we've been on the sidelines for some time out here. From now on we'll be in the middle of it."

I hoped they had been inspired by my little speech. I couldn't dwell on their reactions, as we now had to go about the business of anchoring the boat. Ted had trained me well on so many aspects of our work but one activity we had never spent much time doing was actually setting up the boat for digging. This required precise positioning of the boat and its multiple anchors, made even more difficult by the strong current and lack of sand for the anchors to

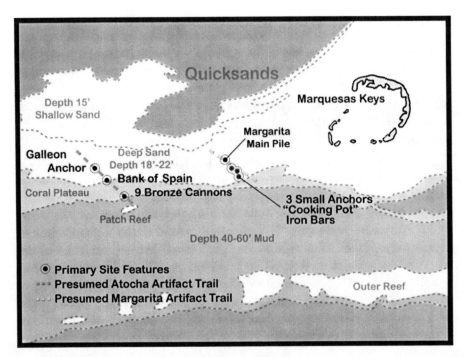

hold. Ted had rarely set up the *Swordfish* while I was on board, and had never discussed his technique.

The tide changed direction about every six and a half hours, moving in a river-like current across these shallows into the Gulf of Mexico or out of it. The only way I knew how to anchor the boat was with the bow facing into the current. To control our lateral movement I ran out both stern anchors almost directly out on each side. The boat then could be winched back and forth sideways, pivoting from the bow anchor. We could move fore and aft by letting off or winching in on the bowline.

This simplistic method was workable until the tide changed. The angles of the stern anchors usually wouldn't hold the *Swordfish* against a strong, stern-on current and the boat would drag out of position. When that happened, we had to raise the mailboxes, recover all the lines and anchors, and then begin the setup process all over again into the current. The procedure was slow and clumsy due to our lack of experience. We also had to be especially careful not to drag anchors or heavy ropes over the bottom near the wood structure area.

All things considered, my first trip as captain went well enough. I managed to put the boat pretty much where it needed to be. Nobody got hurt. The weather and sea conditions were good in between isolated squalls. We didn't find anything flashy, but we did discover some new artifact areas and recovered a broad spectrum of artifact types. Every evening I managed to tiptoe the *Swordfish* through the shallows just off the Marquesas for our nightly anchorage without running aground. The crew had granted me a honeymoon period free of hassles or challenges to my authority.

There were some unexpected new problems. Since the ship's thirsty main engines ran constantly while we used the mailboxes, our fuel consumption now limited the length of the trip rather than the lack of diving air. The anchors also proved ineffective. The *Swordfish* was equipped with a size and type of anchor intended for use in deep sand or mud as on the *Atocha* site. Each one weighed about one hundred pounds with a ten-foot section of chain that attached them to the anchor line. On the *Margarita* site, the anchor's flukes dragged through the shallow sand, often snagging on solution holes where they bent out of shape. Most troubling of all to me were site-mapping difficulties. The LORAN system had worked for general navigation and survey duties, but was frustratingly imprecise for plotting our close-coupled excavations and finds.

My experiment in having the *Whaler* tow a diver over the sea floor hadn't produced anything significant other than a few scattered ballast stones found further east and considerably north of the **Margarita Main Pile**. Although nothing else was visible in the area, I decided to mark the spot for future reference with two spar buoys. Each was constructed of a fifteen-foot length of PVC pipe sealed on both ends. These buoys floated vertically due to one end of the buoyant pipe being tied to a heavy anchor so the free end protruded about six feet above the ocean's surface. They were visible for some distance.

A new dynamic was added to the site before our next trip. As the gossip predicted, Momo came back to Salvors to run the *Virgalona*. I met him briefly in Key West. He was middle aged, short, with close-cropped hair and a white flash of teeth that contrasted to his dark skin. Since English wasn't his mother tongue, his communication sometimes included exotic phrases or made-up word combinations. The *Virgalona* crew seemed impressed with their new captain, parroting Treasure Salvors mythology about Momo being "lucky". They left for the *Margarita* one day before we headed back to the site.

When we arrived the next day, Momo had the *Virgalona* positioned right over the **Margarita Main Pile** area with his anchor lines spradled out around it. Leery of snagging his anchors, I decided to explore a safe distance to the north near one of our spar buoys.

Momo's presence implied a competitive challenge. The new site was now shared by two new captains; one with prior experience, the other a rooky. The rooky had the better equipment and knew as much as anyone about the artifact geography of the site. The old hand had years of prior experience on working wrecks of this type. Momo knew the most efficient way to set up and winch the *Virgalona* where it needed to be. I had a lot to learn.

My first lesson came quickly. I had decided to extend our trip to the site by conserving our diesel fuel. The sand in the area where we worked was very shallow, so rather than using the mailboxes constantly I decided to put my faith completely in our new, sensitive metal detectors.

We dropped buoys in a pattern through the area in which I was interested. I sent teams of detector-equipped divers searching around each buoy in an outward spiral out to a distance of fifty feet. In theory the outer boundaries of these circle searches overlapped each other slightly, eventually covering the entire vicinity. I plotted the coordinates of each buoy drop and indicated the distance and direction of any finds relative to each buoy. We didn't have to keep resetting the boat or moving anchors to work through this relatively large area. The *Whaler* was used to ferry the divers back and forth from the *Swordfish* to each buoy.

Momo didn't know anything about LORAN or plotting. All his knowledge about working wrecks such as this came from his experiences on the *1715* sites. To find artifacts, he had to dig holes and use visual references to position himself. He also didn't venture far from where earlier finds were made as long as objects were being found there. Our experimental technique must have baffled him.

My plan worked better than imagined. Although the current fought the divers, they still located and recovered a number of artifacts and coins around each buoy. We covered a huge area in less than a day. I plotted the finds from the compass bearings and distances from each buoy, and found that we had burned very little fuel in the process. I felt particularly smug.

As I prepared to move to another area, Momo called me on the radio. "Ah, Unit 13 *Swordfish*- this is Unit 7 *Virgalona*. Are you guys done in that spot with all the buoys?"

" Yeah *Virgalona*, we've finished detecting there. We're getting ready to move further north."

" Ok, Ok… so did you find anything there?"

"Roger, artifacts and a fair few coins."

"*Swordfish*, if you finish, leave the buoys. I think I gonna dig there."

There were more than a few snickers aboard the *Swordfish* about that. We confidently moved off to our new area while Momo immediately moved the *Virgalona* into the middle of the buoy field. They set up and began to dig while we continued further afield with our circle search method. We considered the area where they dug worked out. Our new detectors had proven themselves.

At the end of the day we were shocked to learn that the *Virgalona* had already recovered about five times the number of artifacts that we had in the same area. Most were found deep in solution holes, apparently beyond the range of our new technology. Old-fashioned simple excavation had done the job. The lesson we learned was that technology remained a useful tool to compliment, but not necessarily eliminate time-proven methods. I never made that mistake again.

The honeymoon with the crew ended as well. Larry Beckman and a recent hiree had an escalating and disruptive dispute. It originally started in port with disagreements on how certain projects should be done. Both were competent individuals, but diametrically opposed to each other's opinions. Their petty squabbling in Key West bloomed at sea, incubated in *Swordfish's* limited spaces.

Their outright push and shove antagonisms above and below water began interrupting the daily activities. It was almost impossible to continually referee them while directing the ship's operations. The attitudes of the rest of the crew were affected by the strife. As captain, it was up to me to solve the problem definitively.

Ted had always hired and fired the employees. He had said nothing to me about my having authority to dismiss troublesome crewmembers. Firing someone in the middle of a trip forty miles out to sea didn't seem a great option anyway. If the troublemaker previously had a bad attitude, sitting around with nothing to do but brood in the heat guaranteed additional trouble. That night we got underway to Key West.

Upon our arrival in port, I called Ted on the phone and told him about my crew issues. His only comment was "Sounds like you've got a management problem." With that I went back to the *Swordfish* to consider my options. My personal scales tilted in Larry's direction. He was senior to the other diver, technically more sophisticated, and didn't drink. The decision was made.

Early the next morning we got back underway to the site. Firing a crewmember wasn't pleasant, but it brought immediate change to the rest of the diver's attitudes. They were happy to have the disruption behind them and anxious to get back to work.

The trip started well enough but soon squally weather set in. Our crucial back deck shade canopy was torn away in a violent storm. Now there would be no escaping the harsh breath of the sun. For a few days it stayed too rough to dig. We broiled in the lee of the Marquesas, exploring the islands with the *Whaler* until the weather improved. The calmed seas saw the *Virgalona* arrive from Key West. They had a film crew on board from National Geographic and were going to anchor over the **Margarita** **Main Pile**. The film crew would document Momo and crew as they recovered artifacts around the structure area. We weren't included in their plans.

With the *Virgalona* hunched over the **main pile**, we would have to work elsewhere. I had been following a dwindling trail of ballast and artifacts to the north that had continued further than I had expected. Everyone on board knew that the main track of artifact scatter led to the northwest from the **Margarita Main Pile**. Since our finds to the north diminished the further we went, some of the crew suggested that we go back to the northwesterly track. Most of what we were finding consisted of just a few small ballast stones, fragments of barrel hoops and scraps of lead sheathing that had once protected the *Margarita's* lower hull from woodworms.

Stubbornly, I decided to continue at least one more day working to the north. We set up the *Swordfish* almost 300 meters from the **main pile** area, further north than we had previously been. It was a long shot, something I might not have done if I had been in a better frame of mind. Everyone on board was moody and felt excluded from the spotlight shining on the *Virgalona* crew.

The day got even more frustrating as we continually dragged out of position when our anchor's grip failed. The sand was very shallow here, with few solution holes in the flat bedrock. The anchors seemed to grab at first, but as soon as we lowered the mailboxes into position the additional drag against the current overloaded their fragile grasp. Most of the morning and early afternoon was wasted trying to establish a foothold against the current's whims. Without the protection of the rear deck canopy against the sun, it was like trying to work in a welder's arc.

After many tries, the anchors finally held and we put the mailboxes down. We gently dusted away what little sand there was, making a series of excavations that the divers would explore visually and with the detectors. When each "hole" was finished we winched the *Swordfish* fifteen feet over and dusted again.

As we winched more to the east, the anchor holding us in that direction broke loose again. I asked Larry to swim down that anchor line and see if he could find some bottom feature that could be used to "set" the dragging anchor. He was only gone a minute when he surfaced less than fifty feet away, waving and shouting as while he hung onto the anchor line.

None of us heard what he said over the engine's rumble. I guessed he had some sort of trouble and quickly ordered Wally to get the *Whaler* ready for a rescue. A surface swimmer would be carried away by the tide. After I shut down the *Swordfish's* engines we heard that he wanted a buoy brought over to him. Wally was dispatched with the small boat and was there in seconds. Larry grabbed the free end of the buoy line from Wally and disappeared again underwater. We watched as the buoy's Styrofoam float briefly spasmed on the surface, and then stabilized. Minutes later Larry was back aboard the *Swordfish*. He panted not only from his exertions but also from excitement.

We had found a bronze cannon. Larry had seen its dark shape on the bottom from just under the surface. If the anchor hadn't dragged we would have winched right over it in the next few holes. The black mood aboard the *Swordfish* was replaced by absolute euphoria. All the hassles, sunburn and empty holes were forgotten. Dirk Fisher had found the last bronze cannons on the *Atocha* five years earlier.

After a quick backslapping session we quietly pulled up the mailboxes and recovered the anchors. I re-anchored the *Swordfish* directly over the top of the buoy Larry had tied to the cannon. This time nothing went wrong. I shut down the engines and all of us quickly put on our gear to inspect the prize.

Completely exposed on the bedrock, it was an imposing presence. The sand around it had been scoured away by the current. It looked huge, easily over four thousand pounds. There was a little concretion on the breech end as well as baroque dolphin handles. The rest of its form was unfettered with sea growth, in fact parts of it looked almost polished. It was easily the most impressive thing I had ever seen underwater. Five feet away lay an encrusted arquebus, a period matchlock musket. We tied a second, much more substantial buoy and line to the cannon and returned upward to the *Swordfish*.

It was time for a little gloating. I went up to the wheelhouse and grabbed the radio microphone. "Unit 7, Unit 7 *Virgalona*, this is Unit 13 *Swordfish*."

Momo answered, "Yeah, this is Seven.... go ahead."

I tried to sound as nonchalant as possible. "Hey, tell those National Geographic guys that if they get tired of filming you finding pottery shards and barrel hoop fragments that they might want to come over here and see our cannon."

"What you say *Swordfish*?"

"I said bronze cannon."

I gently remove concretions from the cannon's breach area as the feature is photographed on July 6th, 1980. Photo: Pat Clyne

In less than ten minutes the whole *Virgalona* crew arrived alongside the *Swordfish* in their *Whaler* to view the find. Iron man Kevin Farr was so anxious to see the cannon that he had quickly screwed a regulator onto a SCUBA tank with no backpack, put it under his arm, jumped in the water and swam from the *Virgalona* underwater with no fins. Fortunately the current ran in his favor. We all dove again together and Pat Clyne, who had accompanied the National Geographic crew, shot photos of me as I gently taped the concretion off the cannon's dolphin handles. With the handles now clean for the first time in centuries, I rubbed them for luck, just as I had on the *Atocha* cannon in Key West.

The light softened in the failing daylight. We decided to retire to our normal evening anchorage off the Marquesas. The Geographic crew stayed aboard a chartered motor yacht in deference to the *Virgalona's* minimalist living quarters. They decided to transfer to the *Swordfish* the following morning to reenact the finding of the cannon. The *Virgalona* also anchored nearby for the night and would get underway for Key West at dawn. Momo didn't like to navigate at night.

Once anchored in the quiet lee of the Marquesas, the crews visited aboard each other's boats. Most of the interplay tonight was on the back deck of the *Swordfish*. Don Kincaid, who had also worked with the Geographic team, had come aboard with Momo and Pat Clyne. I was up in the wheelhouse, writing my entry in the day's log. Suddenly all three entered through the opposite wheelhouse doors, one on one side and two on the other. The timing was strange; I felt as if I was being cornered.

Pat seemed interested only in taking my picture but Don and Momo had another agenda. Exhaustion from sun exposure and the week's exertions gauzed my thought process, but I did not mistake the meaning of their words.

Don and Momo were not happy at all that I had been made captain of the *Swordfish*. Neither was impressed with Ted or his decision to put me in charge. Some of their friends from the original "real crew" sat on the beach while I was out here in a position they thought I didn't deserve. They were watching my actions very closely and were waiting for the opportunity to replace me with one of their buddies. Momo added his own exclamation point to the declaration. "We're not gonna take any Capitana bullshit!" Capitana was Momo's nickname for Ted.

They retreated as quickly as they had appeared. I was simply too tired to process the threat, though I did mention the encounter later to some of my crew. Today's success had certainly helped my status amongst them and I needed all the allies I could muster. At least the Geographic guys would be on board tomorrow to document our discovery and the *Virgalona* would be enroute to Key West.

We left the Marquesas at dawn to get set up before the motor yacht arrived. The ocean was flat as glass and the water clarity had also improved overnight. As we approached the cannon buoy we could look down from the boat and see its dark shape on the bottom. How many hundreds of times had Mel's search and digging boats passed over this very spot on their way to the *Atocha* site? What else might be found around here?

We had set up again over the cannon by the time the motor yacht arrived with Pat Clyne, Don Kincaid and the Geographic film crew. They all came aboard and shot several reenactments of the find and then continued to film as we made preparations for further exploration. I decided to do a quick preliminary search of the immediate area by bracketing the cannon on four sides with detector circle searches. Other artifacts that could help give perspective to the cannon's location might be found on the mostly bare bedrock. I was mystified how this heavy gun had found its way so far north from the primary northwest track from the pile.

While teams of *Swordfish* divers deployed on the circle searches, others grabbed tanks to free swim around the area. One of them who had decided to join in on the dive was the captain/owner of the motor yacht. He returned shortly to our dive ladder to report on something unusual a little further to the north. I left Wally on deck to keep an eye on things, geared up, grabbed a buoy and jumped in to judge for myself.

We swam north just past the boundary of the circle searches. The sand here had deepened and the current formed it into small dunes. I immediately saw what had caught his attention. One of the dunes had formed like a snow bank over the top of a solid mass. I took a closer look. The mass was cannonballs, fused together by concretion. I set the buoy next to the spot and we swam back to the *Swordfish*.

By now the dive teams had returned from their searches. Although artifacts had been found, I was most intrigued by the cannon ball feature I had just seen. I decided to re-deploy the *Swordfish* on the spot and use the mailboxes to clear away the sand.

We quickly moved the boat into a position over the buoy I had left. For once the anchors held securely and the mailboxes were hurriedly lowered and pinned in place. Not knowing how deep the sand was or what other artifacts might be found, I gave the engines a conservative short and gentle burst of throttle to lightly disperse the overburden. Now it was up to the divers. The anticipation on board was even more oppressive than the sun at its zenith. Everybody jumped over the side, which left only Wally, the film crew and me.

Some of the divers descended into the actual excavation while others investigated the periphery beyond. There were surprises everywhere. Pat Clyne spotted a silver bar partially exposed outside the hole. He found Don Kincaid and motioned him to follow. Stuart followed in their wake. Pat pointed out the bar to his audience and fanned the sand away for a better photograph. Suddenly exposed next to the ingot lay gold money chains of different sizes and lengths. Don and Pat eventually located five of them; Stuart would recover another fifteen.

The divers who worked the excavation also shared a bonanza. The blowers uncovered the rim of a huge solution hole that was filled with silver coins. They resorted to using five-gallon plastic buckets to scoop up the loose pieces-of-eight. Lots of other artifacts were discovered in the immediate vicinity. Numerous encrusted arquebuses, swords and more silver bars lay on the bottom near the pile of cannonballs. There were bronze mortars and pestles, silver plates and silverware. Heavy ballast was strewn across the sea floor. It was another main pile all over again, only better.

On deck we had a difficult time trying to keep up with the volume of documentation and storage of the artifacts. Several times we slowed the recovery process to a more manageable level. When our normal artifact wet storage containers were filled, we resorted to the use of all the galley's pots and pans. For the rest of the day we stayed anchored over that one glorious spot, harvesting history. The National Geographic crew filmed our overexposed world of brash summer sun and reflective water. The last thing they shot that day was the crew throwing me off the bow of the *Swordfish* in celebration of our success.

A National Geographic cameraman films, as recovered artifacts are organized and placed in temporary storage containers.

At sunset the chartered yacht left for Key West taking with it everyone except the *Swordfish* crew. The following day Mel, Deo and many others would come out to the site to inspect our finds but for now it was just us, left alone on the darkening sea. Containers that held the day's recoveries blocked nearly all deck space around the boat. In the nautical twilight we got underway for our Marquesas anchorage.

Posing with some of the gold chains as they were recovered

Stuart came up to the wheelhouse during the hour-long passage to re-examine the gold chains for the hundredth time. It wasn't hard for me to understand his fascination with them. The heavy, almost pure gold looked as if it had just been made. He opened the old attaché case in which I had stored them, fondled the links and felt their weight. After a moment he spoke quietly. "Sweat of the sun, tears of the moon."

I looked up from my log entry. "What?"

"I read once that the Incas called gold the sweat of the sun and silver the tears of the moon. Isn't that a profound description? Today we found them both in quantity.... For years I read and dreamed about finding things like this and now I had come halfway around the world on the chance of doing it. Think of the people who mined and worked with this stuff, how it was passed from one hand to another and eventually ended up on the *Margarita*. It sank and was lost for hundreds of years. Think of all the people and effort involved in looking for it - from the Spanish to us. Today we had found it again and we can hold it, but its journey isn't over. How many more hands will it pass through?"

After we anchored for the night, I was alone in the wheelhouse with my thoughts. As the current softly gurgled around the *Swordfish's* cooling steel flanks, I reflected on the events of the last several days. The discovery of the cannon and the cannon ball area presented to us a new undeniable reality. These major artifacts were deposited during the initial breakup of the ship. Our convenient theory about the northwest tracks of both the *Margarita* and *Atocha* artifacts being from the storm that sank them no longer worked. Perhaps if we learned more about the *Margarita's* wreckage trail, we could find the clues needed to locate the *Atocha's* "**main pile**".

On a personal level there were also changes wrought by the discoveries. By continuing to push along the smallest hint of a trail to the north instead of reworking the easier northwest scatter I had inadvertently changed my own destiny. Don's and Momo's threats were no longer viable against the success that the *Swordfish* had just enjoyed. I would still endure many challenges in the form of broken equipment, storms, nepotism, company insolvency and crew problems, but my position as captain was secure. In the Treasure Salvors world, any discoveries or failures were always attributed directly to the boat's captain. My stock in the company was definitely on the rise. Sleeping that night would be easy.

The cannon ball area became known within the company as the **Cannon Ball Clump**. It would ultimately prove to be the richest area on the *Margarita* site in artifact numbers, types and precious cargo. Its real value eclipsed even that.

Del Norte

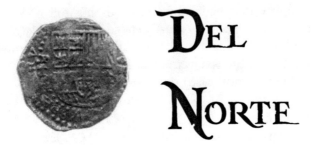

The islands, which constitute the Florida Key's westward thrust into the Gulf of Mexico are few once Key West has been passed. The Marquesas Keys and the distant Dry Tortugas are the last and the most remote features of the island chain. With the exception of a few park rangers on the Tortugas and passing fishing boats, this is an area largely untrammeled by humans. Only the birds and sea creatures registered the passing of time in the form of tides, winds, and natural polyphonies.

On one tiny spot atop the ocean's surface existed a place where noise, heat and human activity were expended well beyond their proportion to the surroundings. High radiant temperatures, as well as laboring machinery below decks, rendered much of the exposed metal structure too hot to touch or walk on with bare feet. A devil's perfume of diesel fuel and combustion were only surpassed by the artillery decibels of large displacement exhaust, insanely rattling compressors and boosted loudspeakers that transmitted radio traffic.

Investigative divers struggled on the sea floor against both tidal flow and the mailboxes' discharge. On deck, coordinates were noted, site maps updated, artifacts tagged and stowed all within the ritual framework of sequential winching of the scary-tight anchor lines for the next excavation. Life aboard the *Swordfish* had become all at once repetitive, physically demanding, uncomfortable and often dangerous.

Continuous exposure to this new, fast paced workload changed many things. We were all deeply tanned by direct and reflected exposure to the sun. The physical demands of swimming hard for long hours underwater, pulling anchor lines against the current and manhandling heavy anchors and chains tempered our muscles. We only returned to Key West to quickly refuel and re-provision, and were back on the site the next day. It was non-stop, seven days a week.

Our operations started at dawn and didn't stop until nine or ten in the evening. At night the still broiling confines of the lower deck bunk areas were too hot for sleep, so everyone dragged their mattresses out on deck. Usually several times every night abrupt and violent rainsqualls passed over our anchorage, forcing everyone back into the discomforts below. As soon as the storm passed, the crew pulled their soggy mattresses outside again in hope of getting a little more sleep until the next rainstorm. Life on the *Swordfish* was hard-core.

The *Swordfish* bakes in the summer sun.

We learned to shed all unnecessary dive gear to decrease drag against the currents. The only thing we wore more of was lead weight for stability on the bottom. This really helped on the occasions we worked under the mailboxes while the boat was excavating. Our fingers were continuously abraded while holding onto the unfriendly seafloor. Under water, blood looked like green cigarette smoke.

No wet suit was necessary in the eighty-six degree summer water but t-shirts helped guard against the chaffing tank straps. We briefly tried surface-supplied breathing air instead of SCUBA tanks but found the long air hoses troublesome against the tide. Our last experiment with it ended while we were digging and my own airline was sucked past the safety cages and into the *Swordfish's* propellers. The hose looped around one of the spinning blades and suddenly I was winched rapidly up into the prop. Fortunately, the airline parted seconds before impact and I was able to swim clear.

Our efficiency improved by leaps and bounds. Ted had finally procured some heavier anchors for us, which usually held and didn't bend. Originally, only Wally operated the *Whaler* when we set up the *Swordfish's* anchors, but I began training all the crewmembers to accomplish the task using an improved method. Our refined technique dramatically reduced setup time. We ran the anchors in or out at the *Whaler's* full speed instead of at idle as Ted had preferred. Setup time with all anchors in place, mailboxes down and excavating was reduced to just minutes. Speed was an ally when we worked in the *Margarita's* strong currents.

My personal skill progressed as well. Having to repeatedly setup the boat four or five times a day, I learned how to position the boat in any direction regardless of the current. I studied the relationship of the tide and wind in the area I wished to excavate and could visualize the precise geometry needed for the placement of the *Swordfish's* anchor lines. Even when an anchor dragged with the mailboxes down I learned how to hold our position until it could be run out again.

The *Swordfish*, with its powerful engines, extreme maneuverability and relatively shallow draft was the perfect tool to handle the *Margarita* site's peculiarities. I was already proud of my little command when Ted and Leah honored me with a navy baseball cap sporting "scrambled eggs" and lettered "M/V SWORDFISH". From then on, I would always touch any gold artifacts we found to the hat's gold braid for luck.

My relationship with the crew evolved as well. The roster often expanded temporarily, but I came to rely on a fixed core group. The demanding nature of our work and the realities of discomfort, low pay and no days off quickly soured most people. A mixed blessing for us was that the individuals who were drawn to Salvors' employ usually had very healthy egos. Keeping them all pointed in the same direction with minimal interpersonal friction was important and sometimes required real effort on my part.

With the freshness of my promotion to captain behind me, I sensed that I was being observed and measured more seriously by the divers. In compensation, I tried hard to speak as much as possible in calm, measured tones. This was tough on the back deck when I had to compete with so much ambient noise. Any time a question was asked of me I made an extra effort to make sure my reply was the best, most decisive-sounding response I could conceive.

My nights at sea were basically sleepless, so I used my bunk time to work out the best way to direct emergency or unusual situations. From my new perspective as ship's master, I pondered the talents and personalities of those on board. They were the first crew under my authority, and my experiences with them shaped my management skills with later employees.

Stuart was the surprise of the group. His dark, movie star looks inaccurately suggested a shallow, pretty boy. Even though he had been a commercial diver who worked on oilrigs in the North Sea and Australia, his knowledge of mechanics and boats was practically nil when Ted hired him. The privileged son of a wealthy banker, Stuart had finished his university degree in England before seeking more tactile adventures. Stuart was probably the most intellectual and well read of any of the crew; he researched the Spanish colonial period thoroughly before arriving at the *Galleon*. He quickly mastered the tricky anchor drills in the *Whaler* and automatically assumed responsibility for maintaining the ship's air compressor. Learning about machinery became his latest passion. He was a tireless worker who often goaded others into further exertions.

Larry had brought more than his own side scan sonar to Salvors. Besides being a good diver, he had a broad mechanical background and had even built his own submersible. Fiberglass work, machining, welding, plumbing and carpentry were all skills in which he was fluent. His fair skin, and premature baldness made him look much older than his years. Larry was intelligent with a quick wit that often turned acidic. He was also very opinionated about everything and was not shy about expressing his views. Surprisingly, in bad weather seasickness was a reoccurring problem for him. He enjoyed Stuart's respect for his talents and their resulting friendship.

John Green loved being at sea. When he had signed on at Salvors he had little real maritime experience and few hours diving since his certification courses. He learned quickly, not only his new vocation's requirements but also those of life out on the ocean. Living aboard ship was about the best thing in the world for John Green. Usually quiet and reflective, he was on rare occasions affected deeply by black moods. Fortunately these events passed quickly. By chance he missed out on the day of discovery of the ***Margarita* Main Pile** by being stuck aboard the *Arbutus*. Since the **Cannonball Clump** discovery on July 7th, he had distinguished himself by finding numerous unique and valuable artifacts by doing extremely patient, detailed examinations of each excavation.

The youngest crewmember, Frank, was deeply tanned and sported a perpetual cigarette. A trained commercial diver, his soft, distinctive drawl confirmed his Southern upbringing. Frank had the least aggressive personality of all the crewmembers and did his job without any emotional hernias or competition. Sadly, his tenure with Salvors was short-lived. One night while we were in port, Frank was on Duval Street in front of the Boat Bar. In a case of mistaken identity, two men jumped out of a car and attacked Frank with pool cues. Although he was severely beaten, he survived. After a long convalescence he left Salvors and Key West for good.

Wally was the oldest and a non-diver. He hated to be introduced to outsiders as the cook. He preferred the more nebulous "crew" job description. Under Ted's command his responsibilities were oriented around galley duties and running the *Swordfish's* anchor lines in or out with the *Whaler*. During extended time in port he wasn't expected to help with the ship's maintenance and spent most of his free time at the local bars, particularly "Two Friends" on Front Street. Ted had always allowed a portion of the grocery money to include beer for trips at sea - a nice reward at the end of a long hot day. In time, Wally boosted the quantity he purchased to cover his own considerable thirst. His personality was usually jovial, especially when assisted with beer and cigarettes.

Wally's pleasant world changed with Ted's departure. We spent far more time at sea with only quick turn-arounds in Key West. The rest of the crew was now trained and better at running out the anchors. Because of the additional workload, the ship and its equipment required more maintenance, which I thought should be shouldered by EVERYONE on board. The divers complained about Wally consuming all the beer and the quality of the food preparation. His enthusiasm for my command and the whole program began to fade not long after we found the **Cannonball Clump**. Between meals at sea he often took our *Whaler* over to the *Virgalona*, spending time with their cook who had similar hobbies.

Word trickled down to the boat crews that the State of Florida was now going to challenge Treasure Salvors over the ownership of the *Santa Margarita* just like they had with the *Atocha*. Since neither Mel nor Ted had issued directives to us to do anything differently, it was an easy choice for us to continue on as if nothing had happened. Our simplistic view was that Salvors had beaten them once, so somehow the company would again prevail. The seaborne operations continued.

Three spar buoys now rode with the waves on the *Margarita* site. They marked the positions of the **Margarita Main Pile**, the **Cannon** and the **Cannonball Clump**. A large distance separated these three features, spaced along a north–south line. The site was similar to the remains of a high-speed airplane crash rather than a concentrated collection of shipwreck debris. In a few areas, the artifacts were found densely clustered together. Mostly they were sparse. As I looked across the water at the spars, I found it hard to imagine the power of the storm that broke up the ship and so badly scattered its remains.

After the discovery of the **Cannonball Clump**, I was especially driven to continue exploration further to the north. We did cursory excavations and worked further and further up the line away from the new finds. At first more coins, personal armament and jewelry were found. Eventually, the further we went, the artifact types began to transition into cooking pots, cookware and ceramics. We pushed northward nearly four hundred meters until the trail thinned dramatically.

I changed direction to the opposite end of the trail as an experiment. Since the artifact scatter was strongly defined now on the north /south axis, was the *Margarita* **Main Pile** really the beginning of the track or merely where we first intersected it? The *Swordfish* was set up just south of the *Margarita* **Main Pile**. As I looked north, I could see that the three spars all lined up.

We excavated in a southerly direction, and clues there revealed the reverse order of the grounding and break up of the *Margarita*. Near the *Margarita* **Main Pile** feature there was much large, heavy ballast that indicated that the lower hull had been breached. Seventy-five meters south the ballast stones were few and small, but huge torn and crumpled sheets of lead sheathing were impacted into the bedrock. As the sinking ship took on water, the hull's protective lead sheathing had been ground off. We continued to excavate southward and in another seventy meters the only sign of the *Margarita's* passing was an occasional spike head fused to the bedrock. Here, she had still been afloat but the swells occasionally slammed her into the bottom. As I looked back at the *Margarita* **Main Pile** spar from this spot, it was easy to visualize the advancement of her destruction.

How did the northwest trail of artifacts that had lead to the discovery of the *Margarita* **Main Pile** fit in? Our research indicated that only weeks after the *Atocha* and the *Margarita* sank, a second and much more powerful hurricane had passed through the area. The Spanish salvor Vargas had guessed that the *Atocha* had been moved by the massive storm from the spot they had originally found it. Could this second hurricane, with winds from a different direction from the first, have created a "secondary scatter" of artifacts from the *Margarita's* remains?

Both *Swordfish* and *Virgalona* explored to the northwest of the new features. Sure enough, there was a strong trail of artifacts and ballast that led away from both the **Cannon** and the **Cannonball Clump** to the northwest. That strongly implied that the parallel direction of the known *Atocha* trail was also "secondary scatter". The *Margarita's* artifact scatter from the first storm indicated that wind and waves moving from south to north had destroyed her. It would seem likely that the *Atocha* had followed the same path...these were tantalizing but still insufficient clues.

Investors inspired by the **Cannonball Clump** finds added more money to Salvors' overfilled coffers. In addition to the purchase and refitting of Kane's new boat, a Hatteras yacht named the *Bookmaker* and a large, twin engined Mako runabout were added to the fleet. The yacht was a posh, high-speed express for Mel's investors to visit the site, and would be captained by Ted. The Mako would be used as a survey boat and run supplies out to the site if necessary. The company quickly purchased Larry's side scan from him and then bought a more expensive, state of the art unit as well that had a built in sub-bottom profiler. A sub-bottom profiler could detect the presence of materials of different densities buried within seafloor sediments. Some money even trickled down to the crews; divers were given raises and now made one hundred dollars a week. As captain I made twenty dollars more.

The huge amount of money available allowed Mel to experiment with pet projects. One of his ideas was to build a huge floating metal detector that could be towed by boat. Mel thought that large, non-ferrous metallic concentrations, such as the forty-seven tons of missing *Atocha* silver, might easily be found with such a device. He had a meeting with Salvors' electronics expert Fay Feild about the feasibility of his idea, basically a vastly scaled up hand-held metal detector. Fay told him in no uncertain terms that it wouldn't work. That was a challenge that Mel couldn't resist.

Mel purchased a huge fiberglass hull that was intended for a houseboat and had it towed to the dock next to the *Galleon's* parking lot. It was then hauled out of the water by several cranes and set on the ground not far from the decaying *Galleon's* hulk. A few returned ex-employees and a number of drifters looking for temporary work were put to the task of ripping out all the wood and metal structure built inside the fiberglass shell. With that completed, a new reinforcing wooden framework would be bonded in place to give the hull strength and a giant electrical wire loop installed within the hull's perimeter. Mel envisioned the hull as a massive floating detector hoop. The device was optimistically nicknamed "The Hot Tub". If nothing else, the parking lot where the crew labored on it was certainly hot.

The most uncomfortable place to be within Treasure Salvors' realm was still the *Arbutus*. The dredge crew somehow managed slow progress toward becoming operational despite the "Devils Island" conditions aboard. We shuttled mail, supplies, groceries and any ice we could spare out to them at the beginning of each trip to the *Margarita* site. Often at the end of the trip we also stopped by to show them what we had found. They seemed amazed at the artifacts we routinely found on the *Margarita*. We were equally amazed that they were still there.

On one trip the *Swordfish* developed a serious leak in a stern compartment. We returned to Key West and hauled the boat out of the water for welding repairs. Before the hull had even stopped dripping, Wally had already summoned a cab bound for his favorite bar. For the next few days everyone hustled to affect the needed repairs and repaint the bottom so we could get back into the water. Wally was a no-show the entire time.

When he finally did appear my rehearsed termination speech wasn't necessary. He quickly gathered up his clothes while stating that both he and the *Virgalona's* cook were quitting Salvors. They had formed a partnership to sell hotdogs from a cart in Old Town. For now, the crew volunteered to take turns preparing meals rather than my hiring another body. The menu improved dramatically.

By the time we returned to the *Arbutus* in August, the dredge crew had finally finished its preparations and was anxious to put their powerful equipment to work. Currently the *Swordfish* was the only vessel in Treasure Salvors' fleet that could move around the *Arbutus* and its five-ton anchors. Mel directed us to position the ship in the vicinity of the **Bank of Spain** so they could excavate the deep sand there. We set their anchors so they could winch themselves around the area as necessary. Salvors' divers were needed aboard to observe the excavation process underwater as well as guide the digging progression. There would be a two-week rotation schedule that included divers from both *Virgalona* and *Swordfish*.

Moving the *Arbutus* and its anchors around became a common event, especially since bad weather often dragged the ship from its desired position. Everyone on board groaned with disgust when we had to stop what we were doing and steam over to the *Atocha* site to untangle the helpless *Arbutus*. We soon became very skilled at quickly repositioning them so we could get back to working the *Margarita*. This newfound ability would pay off in the future.

The **Cannonball Clump** continued to produce remarkable artifacts. Momo had fixated all the *Virgalona's* efforts in that spot while we had explored further north and south. Gold bullion, gold coins and jewelry became almost common finds. The *Virgalona's* new cook, Dick Klaudt, made the best and most unexpected discovery of all. Unlike his predecessor, Dick knew how to dive. He became bored as he sat on deck while the rest of the crew and Momo worked in the excavation below. He grabbed some dive gear and jumped over the side to see what was going on.

The velocity of the tidal flow surprised him. By the time he reached the sea floor he was well down current from the *Virgalona's* excavation. As he pulled himself across the bottom hand over hand while kicking, he made progress against the current back toward the boat. Finally he made it to the rim of the shallow sand berm at the perimeter of the excavation. At that moment an object caught by the force of the rushing tide flipped out of the berm, hitting Dick in the forehead. He caught the item and held it up for examination. It was a solid gold plate of intricate Moorish design.

I established a routine that spent several days each trip exploring new areas around the site before settling down for excavation sessions in the **Cannonball Clump** or **Cannon** areas. The crew was much more enthusiastic about the exploratory work knowing that they too would also get a chance to find things in these rich spots. They weren't disappointed. We also recovered gold bars, jewelry, more gold chain and numerous silver coins. The amount of treasure this wreck held appeared limitless.

One excavation just thirty feet west of the **Cannonball Clump** proved unusual. It was a sunny day and the water's clarity unusually good. Almost as soon as they had gone down to check the freshly dug "hole" the divers returned to the dive ladder to report that the whole bottom was glittering with gold. It was hard to imagine what they talked about until a sample was produced - a hand-full of gold nuggets. A large bag or container of panned gold had come to grief here and was scattered throughout the shallow sand. Many hours would be spent picking up nuggets, which ranged in size from several ounces down to the size of dandruff. Thousands of nuggets would be recovered.

The twenty third of August found us at the end of a trip that had not gone so well. We explored the **Cannon's** secondary scatter area to the northwest with little to show for our efforts. Tempers had gotten especially short lately from the strain of living too close for too long. Our progress was continually interrupted by anchors that dragged and equipment that malfunctioned or broke. Everyone had gotten cranky.

As much as we struggled, Momo and his crew triumphed. They were in the last hours of a very good trip and had found an area that had produced a rare scatter of gold coins. Across the water we could hear their shouts of excitement and laughter. Before noon they were underway to Key West, leaving us alone on the site.

I decided to bridge the unworked area between where they had been working and our most recent excavation. We moved the *Swordfish* near where they had been. I didn't want to start a dig in Momo's holes but rather just beyond them. I asked John Green to jump over the side and see if the ocean floor beneath us was unworked.

He returned to the surface unsure. I sent Stuart down to check. He came back uncertain as well. In frustration, I put on my mask and fins and jumped over the side. After a couple of free dives I found the spot I wanted and instructed Stuart to winch the *Swordfish* over to where I tread water. With that accomplished we were ready to begin our excavations again. As we had no idea how deep the sand was here, I told Stuart to gently engage the engines for just thirty seconds at a low R.P.M. so even the smallest artifacts would not be displaced. I stayed in the water and hung onto the mailbox frames until the brief excavation was completed.

My intention was to quickly free dive to the bottom and check the depth of the sand after the "dig". The second Stuart shut the engines down I took a deep breath and was on my way down before the regular divers jumped in.

The current quickly cleared away the disturbed sediment. I could already see portions of the exposed seafloor during my descent. At the same time, the form of a diver just descending from the dive ladder became visible through the silt fog. The sand on the bottom was shallow, less than a foot deep. The excavation berm framed the now bare and featureless bedrock in front of me. No artifacts were apparent at first, but an isolated dull glow under the shadow of the boat caught my attention. I swam deeper.

The glow became clearer as I kicked toward it. In one pile on the bare bottom was a tangle of gold money chains and gold bars. The sterile mass lay undisturbed, lying roughly in the shape of a rectangle as if it had once been in a box. There was nothing left of whatever its container had been. I touched the naked gold's impossibly clean form and tried to hold my breath a few more seconds. In the quickly rising discomfort just before panic I grabbed up the entire cache and kicked violently off the bottom for the surface.

Inertia carried me up so that my head and hands emerged briefly about two feet above the surface. I caught a quick breath and glimpse of Stuart and John Green's surprise at what I held before being pulled back under by the weight of over twenty-eight pounds of gold. I kicked at my limits to stay up, sputtering to the dive ladder where many hands helped relieve me of my burden. The mood on board was instantly transformed.

We marked the spot where the gold had been found with a buoy, and logged the coordinates. I laid out the find on the deck as it had looked underwater so we could photograph it. In all there were nine gold bars and eight large gold money chains. The divers returned from their search of the bottom, nothing else had been found. Most shipwrecks produce dense concentrations of artifacts whose relationships are measured in centimeters, requiring elaborate grid systems. With the exception of the **Margarita** **Main Pile** and **Cannonball Clump**, the *Margarita's* wreckage trail was a vastly dispersed smear.

I wasn't sure why I had recovered the artifacts rather than letting the working divers have the honor. One crewmember on that dive was a replacement, on board for a short time before he returned to the boatyard to help on Kane's new boat. He was much older than the rest of us, out of shape and had proven to be nearly blind underwater. We had to team one of the regular crew with him in order to find the things he always missed. Perhaps it was my own ego that led me to pick up the treasure before he stumbled into it.

Without realizing it, I had gained a real-life perspective on the difficulties that the Spanish salvage divers had faced hundreds of years ago. Without the modern power of the blower's thrust, my breath-holding ability wouldn't have been enough to allow more than the most remedial hand excavation. The objects the Spanish were able to find must have been exposed, but the broad scatter path of the wreckage and shifting sand hid much from them. It was no longer surprising that despite their years working the site, we were still finding so much.

Nine gold bars and eight gold chains (28 lbs.) recovered in a free dive

The day got better. As we gently dug toward our original position, we located a large silver bar and an ornately carved ivory box lid of far-eastern origins, still attached to delicate silver hinges. Though impossible to tell, we theorized that the lid might have originally been associated with the gold bar/chain cache. The next day, as the *Swordfish* headed back to Key West, I radioed ahead for Leah to "put on her red dress." That was our personal code of celebration when the *Swordfish* found something especially valuable. Leah would meet us at the dock in her red dress quite often while we worked the *Margarita*.

A new Treasure Salvors sub-contractor named Denny Breese brought his compact survey/dive boat *Tern* to Key West. Denny was an experienced wreck salvor with a varied professional marine background; he had even worked as the divemaster for the Peter Benchley movie "The Deep". He spent a great deal of time fabricating and testing a new metal detector, which would be pulled by his boat along the bottom like a sled. It would be able to detect both ferrous and non-ferrous metals, but would have a limited range. While in the process of preparing the device, he conferred with Mel and Ted about improving Salvors' navigation system. There was a new, extremely precise sys-

Stuart shows me the silver bar he found.

tem used by marine surveyors called a Del Norte Trisponder.

Just months earlier the proposal for us to invest in this very expensive equipment would have been ridiculous. We had depended solely on LORAN for positioning ourselves on the sites because it was simple, widely used and we (through Ted's personal funding) could afford it. Now with all the intensive exploration being undertaken on the *Margarita*, LORAN proved far too crude for detailed plotting. Salvors' currently hyperventilating cash flow excesses solved the problem.

The Del Norte equipment required the erection of two towers on which the closed circuit remote transponders would be mounted. Not far to the northeast of the *Margarita* there were already two sturdy steel I-beam pilings intended originally for aerial guidance to a nearby sunken military target ship. By coincidence they were located exactly where we needed them and projected upward to the perfect height above the water for one of the units. The other tower we fabricated out of scrap from the *Galleon's* parking lot. It was erected on the sea floor due west from the two posts, and north of the known *Atocha* trail.

Installing the west Del Norte tower on the sea floor

Ted and I spent a whole weekend on his garage floor generating charts for the new system. The equipment's coordinate display wasn't in latitude and longitude, but rather very exact distances electronically measured from each tower's transponder. The distance could be measured in increments as small as ten centimeters. By knowing the positions of the two towers, we made charts that indicated distances from the towers in ever increasing concentric circles. The chart's distance circles from each tower overlapped. By reading the distances from the ship borne equipment, it was possible to plot an exact position relative to the location of the towers.

Because of my familiarity with the system, the *Swordfish* was the first to try out the new equipment on August 27th. After we placed the remote transponders with their battery power supplies on the towers, we headed for the *Margarita*. The *Swordfish's* new "receiver" antenna was mounted on the stern mailbox support structure, directly over the epicenter of our excavations. We began digging and searching a series of holes while plotting the positions.

The precision was stunning. If the boat moved more than three inches in any direction the coordinate readout changed. It was almost too sensitive, as any seas rocking the boat caused the readout to fluctuate with the antennae's movement. Final confirmation came the next morning. Without any visual references, I was able to set up the boat and position the blowers exactly over a tiny solution hole from coordinates I had recorded the previous day.

Upon returning to Key West I enthusiastically reported to Ted on the successful trials. Not yet apparent to us was that the Del Norte Trisponder was the most important piece of equipment that Salvors would ever own.

A brief but interesting event transpired that season. The King of Spain, Juan Carlos, visited Key West to see the artifacts we were recovering from the 1622 galleons. To provide a private viewing, we used a local bank's conference room and displayed examples of various artifact types. In 1976 Mel had presented the King's wife, Queen Sophia, with one the best of the recently found *Atocha* cannons. The King came to inspect the latest finds.

He was polite, well dressed and looked quite like an ordinary businessman. I watched him wander along the tables asking questions and making small talk. His predecessors had ruled the world, funded by the likes of the treasure laid out before him. As impressive as the collection was, it alone wouldn't have saved the empire. Spain didn't know it at the time but their position as world superpower was sinking just like the *Atocha* and *Margarita*. As a keepsake, Mel gave him a silver piece-of-eight. A rare "family" portrait was then taken in front of one of the tables of all the Treasure Salvors employees and associates.

The King of Spain, Juan Carlos, examines some recovered *Margarita* treasure as Mel Fisher looks on.

425 Caroline Street

Juan de Chaves squinted into the sun's glare as he watched the shivering diver pull himself over the gunwale and into the longboat. As salvage auditor, de Chaves had a special interest in closely watching all the diver's activities. This one had been in the water most of the afternoon and had finally gotten out for a brief rest and warm up.

The diver straddled one of the plank seats, gulping sweet water out of a small earthenware jar in between a few bites of bread. The sun had already started to dry his skin and hair but the coarse and rumpled cut off pantalones he wore were still soaked and dripping.

Juan didn't particularly trust any of the divers. True, they worked very hard but the nature of their underwater activity hid much from him. Heavy or large items weren't the problem; the divers attached ropes to them so that the boat crews could pull them to the surface. It was handling of the smaller things that raised suspicion. Individual coins and small pieces of bullion were often brought up personally and handed up to the boat from the water. Concealment of these types of objects wouldn't be impossible.

As the diver stood up again, de Chaves studied his sodden pants. A telltale geometric shape was just visible under the wet cloth. Juan summoned him and ordered the man to turn out his pockets. Two eight reale coins fell to the deck. After first conferring with Melian, de Chaves recalled all divers to the boat in order to cut off their pockets.

Periodically guests were foisted on us by Mel or Ted, sometimes for several weeks at a time. Often they were allowed aboard the boats as repayment for favors or equipment. Few of these temporary "crew" were wholeheartedly accepted by the regulars. With such restricted living space aboard there was a natural resentment to outsiders adding their own clutter. Inevitably, the normal pace of deck work or diving operations slowed while we tried to accommodate someone unfamiliar with the routine. My biggest concern was for their safety and how they might jeopardize the safety of my crew. Protests about it to the office were ignored. Soon issues materialized that would finally limit the practice.

Unlike the regular crew, the guests had no vested interest in the project. They had been granted a brief opportunity to be a "treasure hunter" and sometimes proved a little too keen to participate. One visitor ruined a rare and valuable artifact by trying to recover it himself despite instructions to the contrary. Another guest diver's mischief was discovered by accident. Stuart had gone down into the crew's quarters to bandage a cut while everyone else was on deck or diving. He spotted a metal Band-Aid box lying amongst the possessions of our latest visitor. Picking up the box, he was immediately aware of its unusual weight. It was jammed full of encrusted silver coins.

Morale problems became more common, not from the unappreciated guests but from the unbroken months of heat and repetitive hard work. Normally enthusiastic crewmembers turned into whining zombies from lack of sleep and physical exhaustion. I wasn't immune to the fatigue either. This condition was an annual event that would always climax during the last impossibly long days of late summer. Our prime dive season lasted until mid-October when the first northern cold fronts finally reached the Keys. Almost instantly the air and water cooled, and associated high winds allowed us some time off in port. It was like a high and persistent fever breaking at last.

An Australian woman named Pam offered some levity. She visited Key West for a few weeks, hoping to rekindle an old romance with Stuart. We were still short a cook, so for the duration of her stay I offered her the position so she could come out to sea with us. She gratefully accepted and did a terrific job.

Pam's real contribution to improving the crew's attitude was her unflagging desire for the perfect tan. For a great portion of her days at sea she wore nothing but sun tan oil, lying on a towel in plain sight of the crew. Pam claimed to be a dedicated nudist, some of us guessed she enjoyed being an exhibitionist. Whatever her reasons, the crew enjoyed her stay and the petty bickering subsided.

There were significant changes at Treasure Salvors' office concurrent with our prolonged time on the site. The company moved from the cramped Simonton Street address to a large, overtly tropical conch house at 425 Caroline

425 Caroline Street

Street, near the Bull and Whistle Bar. The exotic grandeur of the building marked an esthetics apex for the company. Visitors and investors were always impressed with the romance of the place. A Hollywood movie set designer couldn't have come up with a more perfect location for a successful Key West treasure hunting firm.

Meanwhile, archaeologist Duncan Mathewson hired an expanded team of assistants and conservators to preserve and study our finds. Leah

Miguel was hired as head curator. A new employee, Ed Little, was brought in to plot cumulative site data charts from the search boat logs. This development frosted me at first, since I considered myself the company's prime on-site cartographer. As Ed's function became clearer to me I saw him less as a threat and more as someone with whom to share the workload.

Ted also got his feathers briefly ruffled. One of Mel's early associates, Bob Moran, rejoined the company. He had purchased a large fiberglass motor yacht named the *Plus Ultra*. His intention was to use the boat as a dedicated survey vessel and a wreck site operations "office". He pushed Mel to make him the operations manager instead of Ted; he argued that he would spend all his time at sea and could monitor things directly.

Bob's challenge wasn't completely unfounded. Ted admitted that the company's growing fleet required engineering, repair and equipment acquisition that took all of his time in port. Bob and Ted came to terms and agreed to split the job that Ted had once carried entirely on his own shoulders. Bob Moran would oversee the operations on site. Ted would run logistics and engineering support for all company and subcontractor boats in Key West.

The *Arbutus* crew managed to find a few artifacts in the deep *Atocha* sand. The dredge itself certainly pumped a lot of seafloor material through their sluice box system, but significant artifact areas eluded them. They tried to remain positive, either bravely or desperately determined to make their investment in labor and money pay off. Perhaps they were due for a change of luck.

Kane's new boat, the *Dauntless*, was finally completed. Having spent nearly the whole summer working on it in the boat yard, her mostly neophyte crew was anxious to put to sea. They had suffered through the experience, which was no doubt made worse as they heard of discoveries made by the *Swordfish* and *Virgalona*. For most of the *Swordfish* divers, the *Dauntless'* maiden voyage to the *Margarita* site in September was their first opportunity to see her. The *Dauntless* had the same powerful engines as the *Swordfish*, but Ted had engineered a much more efficient winch system for it.

Aboard the *Swordfish* there was a detached sense of pity for the *Dauntless* crew since they had missed out on all the great finds so far on the *Margarita*. That mood changed quickly. During her first trip to the *Margarita* site, *Dauntless* diver R.D. LeClaire recovered a large gold money chain near the **Cannonball Clump**. The discovery was a brash retort - a gauntlet thrown. From that moment on a perpetual, cautiously friendly competition between the two boats began. The wheezing, pre-Cambrian *Virgalona* was now completely outclassed by the two steel dicers. The *Dauntless* and *Swordfish* even shared the drudgery of shifting the deadweight *Arbutus* into new positions, one trying to do it quicker than the other.

Mel's "Hot Tub" experimental floating giant metal detector was fi-

Kane's new command, the *Dauntless*

nally ready for its big test. The *Virgalona* was burdened with slowly towing the awkward barge-like contraption out to the **Quicksands**. As was his nature, Mel optimistically decided that the device should be taken directly out to the wrecks, rather than wasting a moment testing it first in Key West. He had grandly predicted great success to the media and investors alike.

The sea trial did prove to be definitive. Even though many thousands of man-hours and dollars had been expended building it, Fay Feild's long held opinion was vindicated. It didn't work. It couldn't be made to work. The *Virgalona* wearily dragged it back to Key West.

The "Hot Tub" spent its remaining career with Salvors as an outsized artifact storage tank. Eventually it was sold to someone who wanted to convert it back into the hull of a homemade houseboat. Since Mel never dwelled on failures, the whole thing was quickly forgotten.

After a long trip at sea I especially looked forward to the established tradition of visiting with Ted and Leah over dinner. It was great to get away from the crew and the ship's responsibilities, if only for a few hours. In their own distinctly individual ways Ted and Leah had become my own private confidants, eager to share in my successes and sympathetic with problems. We would laugh over gossip sessions about the inbred world of Treasure Salvors' company politics. I was charmed by Leah's endearing personality and I could even sense a genuinely reciprocal friendship developing with Ted.

My lack of female companionship was a never-ending concern for Leah. Bitter, pessimistic waitresses and scarecrow shrimper women were the only females I had encountered in Key West. The nature of our work left little time to seek out interesting girls. A nice looking television producer had once palmed me a note containing her phone number after several days of filming us working the *Margarita*. The gesture was wasted, she lived one hundred and fifty miles away in Miami and I was always out at sea. In resignation, I tossed the piece of paper over the side.

Leah knew that for professional reasons I preferred to spend my recreational time away from the crew. Since I didn't know anyone outside the company's realm, she decided that a friend's daughter might be the perfect solution. With subtle determination she negotiated a blind date.

Alice Green was the daughter of a retired Air Force Colonel and his wife who lived in Key Haven, not far from the Miguels. She was young, attractive and very high spirited, her petite form crowned by an impossibly thick mane of blue-black hair. Though she was not originally from Key West, she had lived there long enough to know the Keys as well as any local. Alice reveled in jumping waves with the family speedboat, darkroom photography and frequent laughing. She was just finishing up the last stages of training for her pilot's license, an interest she intended to make into a career. I felt dull and conversationally clumsy around her at first meeting, but she surprised me by offering to spend more time with me.

My off time soon was divided between visits to the Greens and the Miguels. Alice's parents were also investors in Treasure Salvors, so my knowledge of the company's activities was of great interest to them. Colonel Green, a published expert in antique silver, enjoyed talking about the silver artifacts we found. The Greens didn't seem to mind that I was ten years older than their daughter.

It was nice to have something else to think about other than my duties aboard the *Swordfish*. For many months I had thought of nothing other than trying to live up to my position and the forensic analysis of the *Margarita* site. Free time spent with Alice was pleasantly non-directional unburdened fun. My inner self, which over the year had worked itself into a clenched fist, relaxed a bit with each hour we spent together. My shipboard perspective and judgment felt sharper for the experience. Alice's strong interest in aviation inspired me to start my flying lessons again when I could.

At last, fall arrived and blustery northern cold fronts flicked spray from the agitated sea. While the strong winds blew there was time for a relaxed schedule in port. I indulged the crew with well deserved time-off periods. For the next five months our trips to the site would be limited in duration and much less frequent due to a regular progression of numerous cold fronts.

With bad weather as a legitimate excuse, I took the opportunity to fly back to Wisconsin to visit my parents for a few days. It had been more than a year since I had seen them last and it would be a treat to have a few consecutive days away from the boat. Bragging a little about my adventures first hand would be fun as well.

One evening at my parent's house, an old family acquaintance stopped by. I hadn't seen Katie Budde for many years. Almost at first glance there was a strange and instant mutual affinity, a feeling we had never previously shared.

She and I had grown up along parallel paths. Our families belonged to the same church. We had each dated the other's friends in high school. There had been some irregular interaction between us but never anything romantic. After high school we had gone different ways. There was no further contact nor the desire for any.

I wasn't looking for an expanded love life; my relationship with Alice was fresh and interesting. Despite that, it was hard to deny the strong immediate attraction I felt for Katie. The feeling was obviously reciprocated. We saw each other often during the days of my visit.

Katie was a special education teacher in Milwaukee with a master's degree in education and a new condo not far from my parent's house. Her life was settled and satisfying. She loved her job. Despite the obvious impracticality of the idea, we talked about how we could arrange more time together. I invited her to come and visit me in Key West during her Christmas recess.

The long plane ride back to Key West gave me time to consider the new development. Alice was an enthusiastic companion but would be leaving in less than a year for college to study aeronautical engineering. She was young and very out going, and I knew from my old college days the social opportunities that awaited her in that environment. There was little real chance of my keeping her attention - especially from a distance.

Katie was blonde, pretty, and being my age was beyond the smorgasbord of collegiate dating. We also shared a new and powerful dynamic, an undeniably strong chemistry. The truth was that she lived two thousand miles away and had been quite happy in her life without me. We both shared the desire to expand our relationship, but I couldn't quite see how it would work out. For now, letters and phone calls would have to suffice. When writing her I would usually abbreviate her name into KT. The moniker stuck.

Good weather in Key West greeted me when I returned. The crew was relaxed and in improved spirits from their time off. It was time to get back to work.

To improve efficiency, the company's boats had become diversified. Denny Breese's *Tern* and Bob Moran's *Plus Ultra* were now the prime search boats. Their only function would be to tow the remote sensing electronic equipment to look for the rest of the *Margarita* and *Atocha*. *Swordfish*,

Dauntless and *Virgalona* would be principally diving and recovery vessels; either exploring through excavation or diving on selected "hits" located by the survey boats. The crews aboard the appointed dive boats were very happy to have been relieved of the boredom of magging, although the *Swordfish* would occasionally still engage in small area magnetometer explorations.

Bob Moran picked his crew for the *Plus Ultra* from a roster of available personnel. He was particular about the type of people he wanted and chose carefully. There would be little opportunity for his crew to engage in activities outside the rigid, monotonous world of remote sensing survey work. Intelligence, analytical ability and real patience were required. Tom Ford, Bruce Etchman, Fay Feild and ex-*Galleon* tour guide Jerry Cash were among the crew. Dick Klaudt would eventually be tapped to work aboard the *Tern* with Denny Breese. Their lives at sea were to be spent slowly tracking back and forth from one horizon to the other while staring at printed readouts.

Mel wanted all of his newly organized survey capabilities re-focused on his favorite search area. The *Atocha* site's broad mud plane between **The Patch Reef** and **The Cannons** had already been examined innumerable times by every type of device and technique that had become available to us. Nothing had ever been found there on previous salvage attempts except the same bombs and modern junk. During the winter, all the boats spent many frustrating weeks once more re-examining the area. The search boats located hits and the recovery boats dove them. Some items we "found" had been lo-

Bob Moran's *Plus Ultra* rolls in the swell as it tows the electronic survey gear.

cated more than ten times before, but we had to check every anomaly due to the switch to our new precision navigational format. It was difficult to argue against the diver's bitter complaints.

The one really positive thing that came out of the experience was that we all got used to working together with the Del Norte navigational system. The search boats refined their technique and integrated the anomalies they found with exact positioning. The recovery boats learned how to place buoys exactly on those coordinates so the dive teams could immediately find the targets. There was continuous feedback between the boats regarding the size and location of the indicated anomalies and the actual findings. The whole frustrating project inadvertently improved our overall teamwork and proficiency.

During this period, the *Swordfish* was equipped with a new piece of equipment unique to the company. The company bought a high volume compressor that could be used to power a precision diver-operated excavating device called an airlift. Compressed air was pumped via hose to the bottom of a vertically oriented, twenty-foot length of large diameter PVC pipe controlled by a diver. The air rose upward in the pipe, creating a powerful but localized suction at its lower end. It was great for excavating individual items out of the dense, clay like mud.

I was particularly proud of the installation of the compressor equipment, since Ted allowed me to engineer the whole setup unassisted. It all worked perfectly, and the equipment added another dimension to the *Swordfish's* abilities. Ted's lengthy engineering tutelage had paid off.

Christmas arrived and so did KT. True to her word she came down to visit when school let out for the holiday. Not wanting to complicate things anymore than they already were, I told Alice that a female friend from back home was coming to see me for a week. She painlessly accepted the news and told me to let her know when I was again available.

Although I wrote KT about Key West's beautiful warm weather, she walked off the plane to find near gale winds and temperatures in the upper forties. Not sure of how she would react to my world, I was curious to see if the reunion would be as memorable as our last meeting.

The week passed as quickly as a tropical sunset. Every moment between us heightened the attraction, the chemistry still worked at this latitude. I couldn't claim to be the only newfound interest in her life. The ocean's omnipotence inspired her to pursue a diving certification when she returned to Wisconsin. We had talked at length about her spending her upcoming summer vacation diving with us on the *Swordfish*. I wasn't quite sure if it was me or the chance to do something interesting that would bring her back. It was still unclear how…or IF we were going to get beyond a few visits during the year.

With the Christmas break now over, the crew quickly returned to a routine at Salvors. Though the *Swordfish* was now exempt from most survey duties, periodically we still towed the mag to examine our personal areas of interest. With the Del Norte system controlling our course, we blocked out an area not far south from the **Margarita** **Main Pile**. Such precise control of our search pattern enabled us to find things in the same area that had been missed while using LORAN.

A strong but isolated hit was recorded. We stopped to examine the contact and the divers returned quickly to the dive ladder to report on what they had found. It was a large grapnel hook, about six feet in length.

Grapnels that size were like seventeenth century magnetometers. Salvors of that time would drag them along the bottom, hoping to snag the wreckage for which they searched. Sometimes, instead of finding the wreck, they would find the grapnel snagged on a natural bottom feature. If the relatively fragile hemp rope attached to the grapnel snapped, the hook could be lost. Our particular hook wasn't caught on anything; it was just lying on the otherwise featureless bottom. There was no clue as to why it was laying there.

As the day wore on I considered the grapnel hook. It was a salvor's tool, as viable to the original search as our electronic equipment and mailboxes were to ours. It was the first identifiable artifact absolutely associated with the Spanish salvage effort. As I touched its rough surface, it was as if Melian himself was looking over my shoulder. Nearby, we later found a smaller broken grapnel.

Periodically I looked east at the short, fuzzy green line of vegetation levitating just above the horizon. It was the same view of the Marquesas Islands that Melian had seen. For whatever reason, the grapnels had been lost or abandoned on this spot, and over three hundred years later we continued where he had left off. It was almost like we searched alongside Melian instead of centuries after him. I would never again visit the site without contemplating the Spanish salvor's efforts.

The grapnels, along with the three small anchors, large copper tub and banded metal rods we found before the **Margarita** **Main Pile** still posed more questions than answers. There was an historic account of the Spanish salvors having to cut the anchor lines on their salvage boats to quickly escape the threat of a marauding Dutch privateer. Perhaps that explained the anchors, but their almost cross–current direction and lengthy distance from the actual wreck didn't fit that hypothesis well.

I watched our modern spar buoys as they stood tall among the anonymous waves. Even with our super-precise positioning system we still used buoys for visual reference or to indicate where extra anchors or equipment had been left on the sea floor. The Spanish salvor's sole method of marking areas or items would have been a system of buoys as well. Their own spar buoys were surplus wooden rigging spars and empty barrels, which would have also been employed as floats. The current would have been just as brutal in those days and their buoys would have had to be firmly anchored in order to hold them fast. We used concrete blocks as weights; the PVC spars each required eight of them. The Spanish would have had to use whatever heavy weights they had at hand.

I theorized that perhaps similar to our situation, the small anchors might have become surplus to Melian's needs as his own search operations progressed westward. Perhaps they had marked an area explored early on and were left for a later retrieval, which never came. The heavy banded metal bars would have made a secure anchor for a buoy, and since one of them was found right next to the copper cauldron (or diving bell), its buoy may have marked the spot where that item had been left. The broken grapnel would still be practical as a buoy weight, marking the southerly approach to the wreck. Although there weren't enough clues to prove any of this, I found that it fun to imagine the possibilities.

Even with the half-serious competition between the boats, I currently enjoyed a good working relationship with both Kane and Momo. Whatever previously held grievances Momo had against me were apparently past. He still treated the *Margarita* site much like the Spaniards had; he relied on buoys and his own internal compass to find his way around it. He was fortunate to have the services of divers J.C. Borum, Mike Clem and Rick Ingerson to log and plot their Del Norte fixes.

Kane adapted to the new technology in his own way; he duly recorded positions of excavations and artifacts but was guided largely by the instincts of experience rather than charts. My own movements around the site at this time were nearly always based on data interpretation from plotted artifact finds. It was an interesting spread of techniques.

Success on the *Margarita* transformed Treasure Salvors. Funding inspired by the discovery of the ***Margarita* Main Pile** and the even richer **Cannonball Clump,** allowed us to indulge in state of the art equipment, more boats and crew. Anything we needed for the operation was immediately purchased. The company went through money like thirsty man gulps water, and there seemed to be no end to the supply. Search and recovery efforts really spooled up.

The consistency of finds on the *Margarita* spoiled us badly. Every trip to the site produced interesting and valuable artifacts. In addition to pottery, weapons, religious items, galley equipment, rigging and structural parts we almost always found silver coins and some form of gold. It became routine, and the unceasing rate of recovery paralleled the inflow of investor monies.

When not examining anomalies generated by the *Tern's* or *Plus Ultra's*

Gold of all descriptions continued to be found in impressive amounts on the *Margarita* site. Photo: KT Budde-Jones

detectors, the recovery vessels gradually worked through the established areas on the *Margarita*. The site had become large enough for each boat to operate without interfering with the others. Sometime when one location turned out to be particularly rich in artifacts we all gravitated to that spot, overlapping anchor lines to get as close in as possible.

On one such occasion, both the *Dauntless* and *Virgalona* anchored and worked right where I wanted to be. In frustration, I set up the *Swordfish* in a random spot we had never been before. This "wildcatting" was completely out of character for me, it was something I never did. New diver Chuck Thrasher returned from investigating our first excavation with a number of silver coins and a gold four-escudo coin. It proved to be the largest gold coin ever found on the *Margarita* or the *Atocha*. Mel would no doubt have approved of my technique.

Although gold items were fun to find and usually got the most attention from the press or the office staff, the unique nature of many of the artifacts made them more memorable. A simple bronze Madonna and child statuette reminded us that not just the wealthy had made their passage aboard the ships. Intricate "nesting weights" from a scale was found still in their well-made holder, complete with maker's touch mark. Lead musket balls sometimes bore the imprint of human bite marks on them. Each find told us a story.

The often-unsettled winter weather still gave us frequent downtimes in port. With Alice's prompting, I forced myself to use those opportunities for more flying lessons. Ground school was my next big hurdle. I learned of an accelerated seminar in Ft. Lauderdale that would allow me to complete the studies and examination all in one numbing, three-day weekend. I had the money for the course, but lodging and transportation were another matter. My small-bore motorcycle that had replaced the ailing Fiat was under qualified for travels beyond Key West.

Ted announced one day during the summer that he and Leah had business to attend to in Ft. Lauderdale, coincidentally on the same weekend as the ground school class. He suggested that I go with them and we all stay at a motel near where the seminar was being held. It was a practical solution to my problem and I gratefully accepted the offer.

Swordfish and *Dauntless* excavating on the *Margarita* site

The long ground school weekend passed in a blur of concentrated studies. I successfully completed the course and examination. During the long ride back to Key West in the Miguel's car, my now happily unburdened brain finally figured out what had just happened.

Although Leah had always shown a broad interest in me, Ted and I had only conversed on Treasure Salvors subjects. He loudly dismissed my interest in flying as a waste of time and money, often commenting that only Navy aviators were real pilots. It had seemed odd that during our weekend in Ft. Lauderdale, Ted had insisted on paying for my motel room and had refused my offer of gas money. I also couldn't get any definitive information from either of them about the nature of the pressing "business". Their generosity was remarkable and was my first indication that Ted considered me someone more than a reliable subordinate.

The weather moderated as spring approached. *Tern* began to survey the entirety of the known *Margarita* site and extended its search patterns beyond the *Swordfish's* northernmost exploratory excavations. Their "sea sled" detector was only about three feet wide which necessitated each of their passes be only one meter apart from the last. It was challenging work. The *Plus Ultra* began mag and sidescan surveys in the deep mud area, working southeast from the **Bronze Cannons** of the *Atocha*. Progress was periodically frustrated by passing fishermen who would shoot guns at the Del Norte equipment on the remote towers, damaging the expensive electronics and halting the work until repairs could be made. We began to closely watch any vessels near the towers.

The *Tern* chugged along slowly, pulling its detector sled along the invisible Del Norte grid. About 400 meters due north of the **Cannonball Clump,** its progress abruptly halted. The detector had physically snagged on something, anchoring the lightweight boat to the spot. Dick Klaudt put on his diving gear to free the sled and was quite surprised by the problem. A bronze cannon lying next to a large stream anchor had hooked the device. For all the

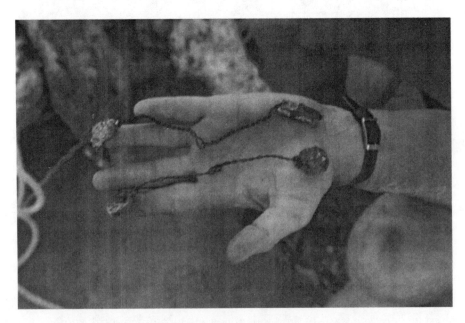

These copper wired "split shot" were intended to be fired from an arquebus (musket).

modern sophistication of our remote sensing equipment, the *Tern* had made its most important find in the same way of an old fashion grapnel hook.

The *Swordfish* and *Dauntless* were the first to begin to explore the spot. We expected another bullion rich area like the **Main Pile** or **Cannonball Clump**, but this area had a different composition. A few silver bars and coins were found just south of the new cannon/anchor feature but the majority of the artifacts were more utilitarian. Galley ware and lots of broken ceramics were the predominate finds. A few marevidis, a type of low value copper coin, were found along with a lead Papal seal once clamped to a paper document. The secondary scatter to the northwest from this new feature was especially strong. One of the most curious discoveries here were numerous copper fragments of rivets and sheet metal, perhaps somehow related to the large, suspected "diving bell" first found a year ago by the *Castilian*.

This latest find was called the **North Cannon/Anchor**. In archival information found by Dr. Lyon, some *Margarita* survivors reported that the ship had broken into three pieces. The ***Margarita* Main Pile**, **Cannonball Clump**

Dick Klaudt (left) and Denny Breese, captain of the *Tern*

and **North Cannon/Anchor** areas certainly fit that description. We continued to widen the search area but there was a feeling amongst some of the operations personnel that the *Margarita's* boundaries had just been defined.

The scope of the secondary scatter leading from the **North Cannon/Anchor** was amazing. It inspired us to look further away from the other main features. Sure enough, we located yet another silver bar and more heavy ballast quite some distance northwest from the **Cannonball Clump**. The power of the second storm must have been quite awesome for it to move the 80 lb. ingot across the sea floor from where it was first deposited. If the material found so far on the *Atocha* was also a result of this same hurricane, how far away might the missing "**main pile**" be?

The *Swordfish* crew was still taking turns cooking while out at sea but there was no longer any enthusiasm for it. By chance, a young woman named Debbie Smith happened by the docks looking for a cooking job. After a brief interview, she was hired to not only cook but to help clean the living areas. She was not a diver, so I wondered how the tedium of being at sea would wear on her. Even though the job didn't pay much, she was still keen to try it. The divers were all very enthusiastic about the idea of again having a cook aboard. Having to constantly stop their regular work to prepare meals and clean up only made their days seem longer.

For partially selfish reasons I decided to appoint a first mate. Being

The lead "Papal" seal (Pope Gregory XIII)

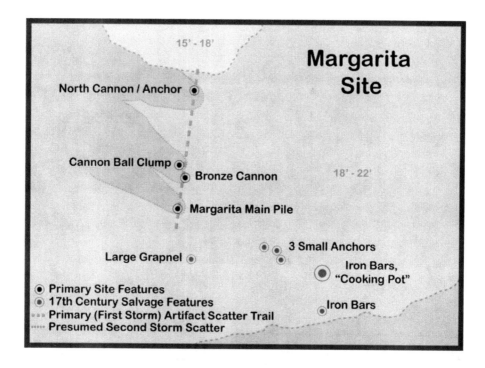

the only person on board with any authority pretty much led to me being on call 24 hours a day. It would be nice to have some help in overseeing the many tasks aboard ship both at sea and in port. Someone in that position could also eventually be trained as a stand in for me if I ever decided to take a vacation. I also thought it might inspire other crew members if they saw that promotional opportunities were available within the company. I repeatedly worked through my various options but one candidate, Stuart, always stuck out.

Of all the *Swordfish* crew, Stuart was the most single minded about Spanish shipwreck salvage. He had read volumes about the subject and it consumed him. The crew all respected his diving expertise. He already called to task anyone who didn't pull his or her own weight. Stuart had a raunchy, good sense of humor that everyone enjoyed and was quickly learning his way around the ship's machinery. Every day he was happy to be in the thick of it, always interested in learning new ship handling techniques and dead loyal to the *Swordfish* and its crew. His appointment caused few problems.

1981 *Swordfish* crew, left to right (upper) Dennis Butler, Chuck Thrasher, Larry Beckman (lower) Stuart Preater, Debbie Smith, Syd Jones. The sunken *Galleon* museum is visible in the background. Photo: KT Budde-Jones

Muddy Waters

Summer of 1981 arrived. I knew that I needed to settle the Alice vs. KT issue in a definitive way before KT arrived from Wisconsin. A turn of fate granted me an easy but temporary way out. Alice would be leaving the Keys with her parents for their summer home in New York State and would be gone by the time KT arrived. In the fall Alice would return to Key West just before starting her first year at Embry Riddle, by then KT would be back in Wisconsin. Rather than having to confront the situation directly, I just let things slide.

Stuart took to his new duties and performed well as first officer. He and I now spent much more time conferring with each other about the ship's operations and discussing the *Margarita* site's daily changing dynamics. In our now closer working relationship he made his opinion clear to me; he didn't want any women working on board as divers.

Many of the divers had girlfriends who made occasional trips with us, but everyone understood that these were temporary visits. Debbie was the cook, and by the unique nature of her job wasn't afforded equal status in the minds of most of the crew. I informed everyone that this summer KT would be diving and working with us as if she were a regular employee, even though she wouldn't be getting paid.

Stuart and some of the others felt strongly that a female attempting to work the same job they did would be detrimental to the way the crew interacted - especially if she were the captain's girl friend. They didn't seem wholly convinced by my assurances to the contrary.

KT finally arrived by car on a sunny afternoon, happy to be finished with the long drive and happier still to see me. It was great to have her here at last, but the romantic adventure would sour quickly unless she somehow could be painlessly integrated with the others. I had absolutely no clear idea on how to work this out. Like a normal new employee she was assigned a bunk space aboard the *Swordfish* and shown where to put her clothes and gear.

It didn't take long for me to see how her presence affected some of the divers. One decided to dispute her abilities almost immediately while she worked on an assigned shipboard project. Her own stand-up personality defused the challenge, but it wouldn't be for the last time. I found myself in troubled waters, wanting KT to enjoy her experience as a crewmember but not wanting to cause hard feelings with the others by implying favoritism towards her. Fortunately, she found an immediate friend in Debbie who by KT's example, was also drawn into expanding her own participation in our work.

Prime diving season was again upon us, and the *Swordfish*, along with the other boats, began the marathon 1981 summer effort. We continued to divide our time at sea between digging through unworked areas on the *Margarita* site and diving on detector hits supplied by the *Tern* or *Plus Ultra*. In the midst of our normal operations the *Dauntless* or *Swordfish* were still called upon to reposition the *Arbutus* when her anchors dragged or her crew wanted to try a new spot. Despite almost desperate efforts, the dredge crew still recovered only a meager collection of artifacts. Our own boat crews grew increasingly weary of supporting the *Arbutus*; they saw it only as an unproductive hassle.

Only the line of fluffy clouds above the great current stream appeared to move. These few long days of slow sailing back to Havana had given Melian plenty of time to think. The almost still air and the sun's unpleasantly bright reflection off the water had made this passage even more frustrating to a man who seldom relaxed or engaged in simple daydreaming. Melian's active mind was always formulating new ideas and plans. He was frustrated not to be able to act on them as quickly as they came to mind. The nearly becalmed ship seemed to make time itself a prisoner of the wind's lazy mood. Melian paced along the wooden decks in an automatic reflex.

The summer salvage season of 1629 had produced few new finds and had been once again disrupted by other challengers for La Margarita's riches. The most recent raid on the site was only the latest in what was becoming an escalating problem with aggressive Dutch pirates.

Almost as troubling was the plummeting rate of recovery of the sunken galleon's treasure. Since the season of 1627, the divers had been able to find only a fraction of what the site had first produced. There had been a brief moment of elation that year when a group of divers searching in the deeper water had found some silver ingots thought to be from the Atocha. Since then nothing more had been found in that area. Considering the Margarita's poor showing of late, perhaps it was time to redirect the expedition more toward finding the rest of the great Almiranta.

Above, the ship's lookout announced the sighting of land. An hour later from the weather deck Melian could see for himself Cuba's cloud-frosted coastline emerging on the horizon. Immediately upon getting ashore Melian would have to begin action to claim his portion of the Margarita silver currently impounded in Seville. For his plans to continue he needed both the money and the prestige of the reward as quickly as possible.

The fourth of July brought a company invasion to the Marquesas. Mel always liked a party, and the national holiday was a great excuse to bring the whole expanded Treasure Salvors staff out for an uninhibited private party on the empty beaches. It was a typical druid festival encouraged by intoxicants and facilitated by the company's good fortune and perfect weather. The Marquesas party was the first time that the whole group had been together. There was a faint, but noticeable division amongst the employees. The rapidly growing shore-bound staff was developing its own culture independent from the seaborne operations.

Before the *Margarita*, the company's personnel roster had been so small that everyone felt an equal unity of purpose. The recent personnel expansion left two loosely independent groups who squinted at each other from increasingly distant perspectives. Comments at the party from some of the office staff highlighted their dim understanding of the boat crew's function at sea. One naively assumed that we spent our time in perpetual beach parties like this one.

An unexpected event soon tipped Treasure Salvors on its side again. Bobby Klein, a new competitor for the *Atocha*, appeared and began his own search just outside our Admiralty law claim. He began to survey confidently from a large speedboat in the deep water south of the *Margarita*, and southeast from Salvors' *Atocha* finds. He was careful not to infringe upon our legally protected territory, but his sudden and persistent presence was a surprise to everyone, especially Mel.

It was the Olin Frick affair all over again. With the bulk of the *Atocha's* remains still missing, there was the old possibility that they just might lay outside of our continually explored, claim-protected boundaries. No one knew how the Admiralty court would rule if someone else found the vast majority of what we sought outside the area we had been granted. Mel was in an absolute panic.

Over the years Mel had stubbornly repulsed all officious arrows targeted on him by the Federal and State of Florida governments over ownership and salvage rights to the *Atocha*. He was confident about the pending final rulings for the *Margarita's* ownership against the State of Florida as well as the Bobby Jordan mess. The appearance of this new search boat near our claim border was a sobering manifestation of the worst-case scenario for Mel and Treasure Salvors. Could the ownership of the unfound *Atocha's* majority still be lost despite all the years of effort and suffering? Mel launched an immediate counterattack.

A plan was dispatched to the crews from Bob Moran. Mel wanted all the search boats to blanket survey the area where the new intruder was searching and beyond. If we could find something from either the *Atocha* or *Margarita* in that area before Bobby Klein did, we could have our claim boundaries legally expanded to protect our interests. Because of the serious potential consequences, Mel insisted that the search boats run 24 hours a day, racking up hits and assigning the most promising of them to the recovery vessels for exploration. The *Bookmaker*, captained by Kim Fisher, also joined the fray as a survey boat. Mel's interest in keeping the boat as a pristine yacht had withered quickly, and against Ted's protests had it converted into a workboat.

Originally Mel also wanted the recovery boats to work 24 hours a day as well, arguing against our real safety concerns. Simple diving physics influenced the final ruling; the deeper water required more decompression surface intervals than round the clock diving would allow the available divers. The *Swordfish*, *Dauntless* and *Virgalona* crews would get to sleep at night while the search boats continued to slowly trace a grid of electronic seafloor exploration. It looked as if it would be a long, unusually tedious summer.

Just before this latest company trauma I finished all the requirements for my pilot's license and only had to pass a final flight test with a FAA examiner. I scheduled the test carefully so it would coincide with our next brief reprovision in Key West. Although Mel was anxious for all the boats working to be out on the site as much as possible during this latest crisis, the crew could prepare for our quick return to the site while I took the hour-long test.

As anticipated, the *Swordfish* ran low on water, food, ice and fuel on the day I was scheduled to take my flight test. While we made preparations to get underway to Key West, Mel called on the radio to check on our status. I explained that we were on our way in to reprovision, and would be coming back out as quickly as we could. His unusually frantic response showed how seriously the interloper's presence affected him. He ordered us to stay put; he would charter a boat to bring us the things we needed. I reminded him that there would still be five of his boats on site, but his anxious state of mind allowed no deviation. My long awaited flight test was apparently scrubbed indefinitely.

Ted saved the day. After hearing the radio conversation, he brought his own boat out to ferry me back to Key West so I could make my appointment. Since Stuart wasn't ready yet to take full command in my absence, Ted brought recently rehired Craig Pennington out to fill in as "captain du juor". Thanks to Ted's quick action, I was able to take the test. He returned a freshly licensed pilot to the *Swordfish* the next morning.

The search boats never stopped moving except for seaborne transfers of fuel and supplies. Their crews were divided into watches, never having the luxury of enough sleep. They compiled a list of hits for the recovery boats

Even as a volunteer, KT had to prove herself to the other crew. Here she changes the heavy batteries on the slippery Del Norte west tower.

during the night and every morning radioed us a fresh batch of coordinates to dive on. The tone of their broadcasts grew ever more fatigued as each day passed.

One morning the *Plus Ultra* radioed us the position of a strong but singular mag hit far to the south of where we had previously dove. We found the spot to be just north of the outer reef in water about forty feet deep. KT and I sometimes paired for circle searches, and that morning it was our turn to make the first dive. I anchored the *Swordfish* close to the buoy we had thrown out to mark the hit's indicated position.

The water that morning was unusually clear, and from the surface we could see the marker buoy's concrete block resting on the brightly lit sand bottom. After all the countless circle searches in the shadow-world of the mud zone, this spot would be a real treat. The ocean felt as warm and inviting as a broad sunlit meadow. KT led our descent, trailing a rising curtain of reflective bubble stream. She carried the control line that would be tied to the concrete block. I had the detector and followed about fifteen feet behind her. She swam straight to the buoy block and prepared for the search. About ten feet away from the block was a small, algae-bearded outgrowth that protruded from the lightly rippled sand bottom. It was probably a piece of old coral rubble but I turned the detector on and passed the loop across the mass.

The detector screamed so loudly that KT heard it ten feet away over the noise of her own exhalation. Whatever it was, the detector indicated there was a lot more of it still buried. After a brief attempt at hand fanning the loose sand away from the object, we surfaced and told the crew to prepare the airlift compressor and equipment. In minutes we solved the mystery.

The airlift efficiently exposed the prize. We had found an anchor, about seven feet long. Most of it was deeply buried in the soft sand and the part that had been visible was the very tip of one of the flukes. Lost or abandoned anchors weren't uncommon, but this one was special; it was the correct design for the 1622 fleet period and it was "set" in its working position. Even more remarkable was a large fragment of the original hand-carved wooden crossbar still in place. The orientation of the anchor pointed north, right toward the distant *Margarita*.

This auxiliary or "bower" anchor wasn't big by galleon standards, but its discovery caused a huge spike of activity and adrenaline in the previously fatigued pace of operations. The *Plus Ultra* crew immediately began reviewing their records for any hits between this new anchor and the *Margarita*. All our survey boats began to crisscross the area with new intensity.

A few days later we were called over to help the *Dauntless* dig out another *Plus Ultra* mag hit with which they had been struggling. This spot was some distance north of the anchor the *Swordfish* had just found. The bottom was the more typically silty clay-like mud of the deeper water area. Our airlift was again deployed and with the *Dauntless* crew we dug up the next surprise. We found a large galleon anchor nearly twelve feet long and in excellent condition. Suddenly, searching in the gloomy mud zone had become exciting.

This anchor also pointed toward the *Margarita* and was directly in line from the first one. Our already stepped-up activity level went into hyper drive. The *Dauntless* was soon equipped with its own airlift compressor, and all the boats pressed hard for the next big discovery.

Many hundreds of hits from the survey boats would now be anxiously circle searched. Amongst the old expended military munitions and modern junk we found were two more anchors. They too were in a direct line between the first outer reef anchor and the *Margarita,* and like the others were "set" in the precise direction of the *Margarita's* primary scatter. The smallest of the new pair was a real beauty; the dense mud had perfectly preserved its entire wooden crossbar and there was even a date stamped in the iron shank-1618.

There was overwhelming evidence that they were all from the *Margarita*. After being carried over the reef by a freak wave, the *Margarita* crew had tried to stop her storm-driven progress towards the dangerously shallow **Quicksands**. The ship's inertia had proved too great for the hemp rope tied to the anchors; as soon as they were dropped over the side and had dug into the bottom the lines had snapped. Their loss was our gain as we now

A plume of mud comes to the surface as a *Swordfish* diver digs out a *Plus Ultra* anomaly with an airlift.

had absolute proof of the original storm's wind direction. It showed us definitively that Salvor's limited *Atocha* finds could only be from secondary scatter. If that was the case, where was the primary?

The discoveries soon waned. We followed the survey boat's detector hits all the way north to the **Margarita Main Pile** but nothing else was found along the anchor trail. At least it proved the original theory that the **Margarita Main Pile** was indeed where the ship originally started its breakup. The discovery of the anchors had also resolved the Bobby Klein issue. We were able to legally expand our protected area to include the territory he'd been searching. Salvors had been fortunate to find the artifacts first.

This season brought change at many levels. Crew frustration aboard the *Swordfish* grew inversely to the lack of sustained finds. John Green transferred over to the *Plus Ultra* to help with the survey work. Larry became more caustic, openly challenging my judgment loudly in front of the crew. Was he simply getting burned out or had I lost the bubble of management?

I learned that one strongly dissonant voice could rapidly degrade morale amongst the crew. We all spent far more time with each other than most married couples could stand. After lamely trying to work through Larry's behavior, I arranged for him to transfer to the *Dauntless*. He never settled comfortably there, and soon begged to come back. His complaining had stressed everyone enough, since life onboard was appreciably calmer without his outbursts, I refused his return. He soon made a permanent departure from the company.

Stuart and Larry were such good friends that I half expected a backlash of some sort from my new first mate, but it never came. In time, I would understand that Larry's change of attitude was just a symptom of an incurable frustration with life at Treasure Salvors. Over the years the disease was caught by many others and was usually fatal to continued employment. There came a time when there was no further joy in the long hours or discomfort, and in periods of finding few artifacts the promised "adventure" turned bitter. No one was immune.

Stuart's own issues were more unique. His visa had run out and he was faced with having to leave the country. In the spirit of trying to preserve her *Swordfish* "family," Debbie the cook made a remarkably gracious gesture. She would be happy to marry the Australian so he could stay.

It was an odd paring. Debbie functioned well at her job, but in port was a loud and uninhibited party girl who was drawn to wherever the punk rock music was the loudest. Stuart was as conservative as an English banker and spent his evenings reading. They both had been drawn far enough into the *Atocha* and *Margarita* project to take unconventional actions.

They were to be married by a T.S. company receptionist who hap-

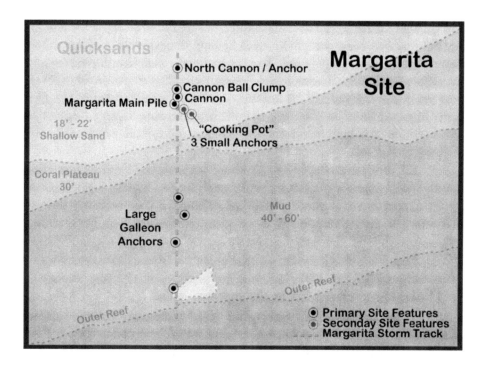

pened to be a notary public. The ceremony was appropriately held aboard the *Swordfish*, and instead of a tuxedo, Stuart wore his dive gear. When the notary told Stuart he could kiss the bride, he offered her his regulator and said, "Suck this." Debbie sported drink umbrellas stuck in her pigtails and flamingo sunglasses. A "Just Married" banner was tied to the mailbox framework and we quickly left the dock for the *Margarita* site towing strings of buoy balls instead of tin cans. The marriage wouldn't last long, but it would keep the whole crew together a while longer.

The dredge project aboard the *Arbutus* wilted at the end of the season. Like so many subcontractors who would try their luck on the *1622 fleet wrecks*, the crew were wholly disappointed in the few artifacts they found for all the money and effort expended. There had been no pot of gold at the end of their borrowed rainbow. Some of the dredge crew settled for a time in Key West, but many just faded away. The *Arbutus* itself returned to its former near-derelict state, manned singly by unlucky T.S. newbies.

It was nearing time for KT to return to Wisconsin when Ted changed everything. Both he and Mel had closely watched her performance during the summer months. With just a few days before her scheduled departure, Ted surprised her with a job offer as a real Treasure Salvors diver.

Those of us who heard the rapidly spreading news were as shocked as she was. Deo Fisher, Angel Fisher, Jo Arden, T.T. Hargraves, Bleth McHaley and Claudia Linzee were all women who had spent time working the *Atocha* site but that had been years before and beyond the experience of any of the current crew. During my tenure, females had never been considered for any operations work other than as cooks. In addition, KT had spoken often about how much she enjoyed her job and new condo back in Wisconsin. She had never inquired about employment options with Treasure Salvors. Ted's offer was a serious turnabout in current unwritten company policy and wildly out of character for Ted.

KT didn't even have to think about her answer. Continuing our romantic involvement was a factor, but by now she had already been seduced by the almost mystical power of the lost galleons. KT immediately flew to Wisconsin to rent her condo and to get a leave of absence from her teaching job.

I used her brief trip home as the opportunity to quickly cash out my relationship with Alice. Alice had done nothing to deserve being "dumped," but I had made my choice. The matter appeared settled.

When Ted hired KT, I was pushed across a boundary into unexplored territory. She was no longer just a guest, but would have to be treated as an employee, and seniority-wise, a very junior one at that. How I treated her relative to the other crew was like walking a tightrope across the center of a minefield; falling to either side would be bad. I hoped our developing bond would survive its new reality.

KT quickly returned from Wisconsin to rejoin the *Swordfish*. There was far less exchange of divers between the boats now, and the intramural competition between the recovery vessels produced an increasing polarization between them. Though still united by a common goal, each boat had almost become like an independent country. Each had its own unique political structure, value system, mythology, and standard of living. Significant to the reason for this expanding drift were the physical differences between the boats and the personalities of the management. How each captain spent his off times was different as well.

Momo was easily the oldest of the three recovery boat captains, and had a family in midstate Florida. For years, though thick and thin he had been a closely held friend of the Fisher family. His accomplishments and valued status with the Fishers left him with little to prove. When bad weather or mechanical problems brought the *Virgalona* into port for several days, the dock lines would barely have been tied before he was in a car on his way to see his wife whom he nicknamed "The Beef". Momo's quick departures usually gave his crew a lot of unsupervised time to casually enjoy themselves in port. Maintenance or preparations that were needed for the next trip happened frantically at the last possible moment before going out again. Momo sometimes had a difficult time remembering his crewmember's names, so he often gave them Spanish nicknames. Of course these nicknames stuck and became permanent within the company.

The *Virgalona's* divers acknowledged their boat's somewhat underdog status by proudly calling themselves "the Black Sheep Squadron". To most of them, their old blue boat assumed the style of a convertible '57 Chevy, not fast or powerful, but vintage cool.

The *Dauntless* boys loved Kane. He was like a direct connection to Mel Fisher, the "World's Greatest Treasure Hunter". Because of that, there was often a pervasive sense of entitlement amongst them. Both Mel and Deo were very fond of all their children, and now it was their youngest son's turn for ascendancy. Like the other Fisher kids, Kane's early childhood years were entwined in the family's salvage efforts on the *1715 Plate Fleet* sites near Vero Beach. He was still quite young when Mel and Deo moved their operation and family to the Keys in the search for the *Atocha*. Kane grew up much like a farm kid, sent to school but expected to help in the family's endeavor. Like his other siblings, he was absorbed into his parent's unusual calling.

Free spirited Mel put few restrictions on his children, allowing them to go through life's experience in an unbuffered fashion. Older brother Kim and younger sister Taffi had easy, outgoing personalities but Kane was often shy around unfamiliar company. Mel readily included his offspring when promoting Treasure Salvors, and Kane's successes aboard the *Dauntless* were usually presented to the media.

Kane was generous to his crew, having the deep-pocket resources to sometimes buy them meals in port or even give gifts of treasure coins. They completely believed in his ability to find things on the two sites, almost as if it were part of his genetic code. Years of being in the company's operations made him an expert in ship handling and diving. It was hard to imagine him doing anything else.

Kane was from an unconventional family, living in an unconventional place, engaged in an unconventional lifestyle. More and more he seemed to spend his off time roaming the lower Keys with a small but regular posse of society's fringe. Despite occasional rumors, none of us really knew what adventures they were up to. He was the embodiment of Key West itself – having the potential of unusual promise, but teetering on the edge.

Though I worked for Mel, I was very much aware of my own power-base with Ted. To let maintenance or the ship's appearance slide would have been a rude payback to my benefactor. While the *Dauntless* or *Virgalona* divers relaxed or partied in Key West, the *Swordfish* crew was usually chipping rust, painting, repairing or preparing. Free time was allowed only after the "to do" list was completed. The side effect of all this was an induced professional pride in their work and in our little ship. They liked the *Swordfish's* sharp, squared-away look and work ethic. We had become the straight arrow boat.

For cheap entertainment in our spare time, KT and I would often explore Key West on bicycles. Because of our extended sea deployments, we usually saw the island only in short increments. Biking through the Old Town streets cost nothing and was a good way to observe Key West slowly reinventing itself.

The island's economy had stalled hard years earlier when the Navy dramatically downsized its presence there. Many of the locals had returned to an old Key West tradition with an updated product line…drug smuggling. Participants in this new growth industry ranged from teenagers to city officials. With hundreds of uninhabited islands on the doorstep of the Caribbean, the relatively unpatrolled waters and airspace of the Keys was the perfect place to put fast boats, low flying airplanes and local knowledge to work.

For years the primary import was pot, and plastic wrapped bales of the stuff floated around the islands from missed aerial drops. A popular t-shirt of the time was a take off on a Greenpeace slogan, "Save the Bales". In time, the shift towards harder drugs would introduce a meaner, more serious demeanor to the vocation.

At the same time, the Old Town area was steadily transforming. A growing gay community bought up the decaying, termite-farm "conch" houses and restored them into elegant guesthouses or tropical estates. This began a slow but steady real estate boom that would eventually consume all the Keys. The city fathers debated plans on how to make Key West an exclusive destination for wealthy tourists, a vision hard to imagine at the time with all the homeless dirtbags, empty storefronts, scruffy shrimpers, and pounding gay bar disco music.

The recovery boats eased back into their normal routine of excavating the *Margarita*. There were still many square miles of ocean bottom to explore, but the further we moved from the main wreck features, the more sparse and fragmented the artifacts became. Each trip recovered hundreds of small pottery shards, pieces of animal bones and barrel hoop fragments that had been scattered widely amidst the shallow sand and coralline rubble. To spice things up and to keep up morale we would usually return to the **Cannonball Clump** or the ***Margarita* Main Pile** for a day or two in hopes of picking up a few coins or other treasure items.

Most of the precious artifacts were still being found close to the north/south line of destruction from the first storm. The company had recently equipped us with a very sensitive new English detector called the Aquapulse, and the divers were now much more skilled at detailed examinations of each hole. Often our visits back where early big finds had been made proved quite successful. One trip KT endeared herself to some of the old hands by finding gold two escudo coins in areas they had just carefully searched.

Even though we became more efficient at working the wreck, the rate of recovery was down significantly from the first year. The *Margarita's* promise hadn't completely faded, but we were all aware that we were working a wreck that had been heavily salvaged by the Spanish and the Dutch not long after it had sunk. No one really knew how much had actually been recovered at that time. Our success so far on the *Margarita* brought us recognition, the funding to invest in first class equipment and had much-improved our search and recovery techniques. Now the company's pendulum began the swing back towards the *Atocha*.

Over the last year Keys residents had grown used to headlines and news reports about our successes, but there had been no real opportunity for the public to actually see the growing collection of recoveries. For a limited time and with great fanfare, we exhibited the *1622 fleet* artifacts at the old East Martello fort near the airport. From the opening until the last day, record numbers of visitors came to see it. Treasure Salvors Inc. was embraced by Key West.

The setting for the display, which ran from August through the winter of 1981, couldn't have been more appropriate. The old fort's brick-arched corridors had long ago softened with age. Dramatic lighting was set up in the timeworn rooms, and the artifacts, both common and precious looked bigger than life. The presentation was humbling and atmospheric, almost like entering a great cathedral for the first time. After dark the mood grew even more intense. It was the first time that all the *Margarita* and *Atocha* material found had been exhibited. Even those of us within the company had never seen the finds displayed all together at one time.

Many Salvors employees were asked to take turns as tour guides, a task the divers enjoyed while in port. The *Swordfish* was undergoing unplanned repairs in the boatyard and it wasn't hard to find volunteers among the crew eager for a break from the hard and dirty work. I found my own time at the exhibit a real ego boost; it was fun answering questions from envious visitors. Surrounded by the magnitude of the finds. I found it hard to comprehend the scope of what we had found in just the last year.

Even with the success of the exhibit, there were a few financial hiccups. Several times over the late summer, paychecks had been held up a day or two. Those of us conditioned by the pre-*Margarita* financial void saw the short lapse as a non-event. A real attention getter was when the *Swordfish* returned from a particularly successful *Margarita* trip. Mel met us at the dock, and after admiring the gold coins, jewelry and other unusual artifacts we had found, announced that he was broke and that I would have to lay off my crew. The work stoppage only lasted a few days before more money was forthcoming, but it was the first indication of money troubles since the finding of the **Margarita Main Pile**.

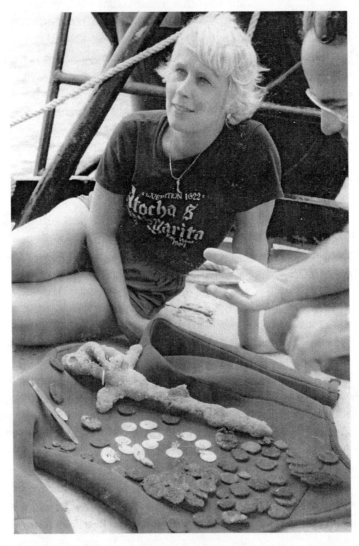

KT and Don Kincaid examine gold coins and jewelry (on white tags), coins, jade and a dagger. Despite the successful trip the company suffered from mounting money problems.

# Ebb Tide

As the *Arbutus* dredge project folded, yet another new subcontractor anxiously prepared to try his luck with the *Atocha*. Ian Koblick brought a large, 150-foot vessel named the *Golden Venture* to prospect in the deep sands. It was equipped with a huge propwash deflector much like the *Jefferson's*, only many times larger. This extremely powerful ship required all the five and seven-ton anchors we could drag to the site to hold it in position.

Unlike the *Arbutus*, the *Golden Venture* had massive winches, air-conditioned cabins and was well equipped to support lengthy diving operations. Once again our divers would be put on a rotating schedule to live aboard the ship and work the excavations.

The bad part was that like all the other subcontractors before them, they would be completely dependant on Treasure Salvors company boats for logistics support as well as for positioning their heavy anchors. The crew of the *Swordfish* and *Dauntless* would often be drawn away from their own work to wrestle the *Golden Venture* and its ground tackle around the *Atocha* site. Instead of waiting for next summer's calm, the *Golden Venture* was going to begin their operations as soon as possible.

KT got her first real taste of new employee status when she drew short straw for *Arbutus* duty. The ship had deteriorated even further, so for safety reasons we now put two people at a time aboard the hulk. With our working boats on the site most of the time, the actual purpose of putting crews aboard the floating wreck was becoming irrelevant. In its advanced decay, it was even more like a prison ship.

We had picked up a few new divers during the summer. The most unlikely one was Bernie, a recent commercial dive school graduate. He was highly prone to extreme seasickness and spent his free time designing unlikely inventions and planning farfetched schemes. He was proud of being the only Jewish diver in Treasure Salvors. Stuart had given him an appropriately irreverent nickname - "Fiddler in the Bilge".

Kim Fisher returned to his family and studies in Michigan at the end of the summer. The *Bookmaker* needed a new captain, someone who knew survey work, the Del Norte system and the sites. Stuart was the obvious choice, so Ted made the promotion official. It would prove a frustrating transition.

The *Bookmaker* had a delicate constitution, more suited to weekend fishing expeditions than daily commercial use. It hadn't worn well during the summer season and everything that could fail was in the process of doing so. Stuart made a few trips, but he and his crew of one struggled constantly to keep the fiberglass beast operational. It overheated, turbos died, the generator repeatedly quit, gearboxes slipped, the steering broke and on and on. When the boat was pulled out of the water for its annual inspection, it was discovered that much of the hull below the waterline had badly blistered from water absorption. The work never ended.

I was happy for Stuart's promotion, but his departure did shift more work onto all of us aboard the *Swordfish*. Around the same time the *Golden Venture* finally arrived on site. With some effort it was set up at Mel's direction near the **Galleon Anchor** area to dig through the deep sands. It began digging huge holes, looking for signs of *Atocha* material. The survey boats also worked, with most of their emphasis on the known *Atocha* corridor and beyond. The *Margarita* was given a breather.

Another new boat and crew had also come to winter in Key West in 1982. While flush with money, Mel had purchased the *Endeavor* for John Brandon to use on the shallow water *1715* sites. Since the winter season was too rough to work on those wrecks, the boat and its crew would normally be idle for those months. Mel decided to have John bring his crew to help work the *1622 fleet* sites until the waters between Ft. Pierce and Sebastian calmed next spring.

Our initial encounter with the *Endeavor* boys was a bit like meeting previously unknown, distant relatives for the first time. The boat itself was small and with a shallow draft, almost toy like by our open ocean standards. It was set up for excavation in very shallow wave wash, just feet off the beach. It was fast, when working the *1715* sites, it could commute daily from dock to wreck site and back instead of staying anchored on site like we did. It had mailboxes like our boats, but *Endeavor's* could be readily converted to a deflector system for very shallow water excavation.

John's crew was barely out of high school. They already moved with a swagger, young roosters confident in their abilities and dead loyal to John. They liked to party, and found that Key West suited them just fine. The *Swordfish's* cook, Debbie, was immediately drawn to the *Endeavor* crew, nicknaming them "The Incredible Edibles". Her real interest was in John, with whom she spent much time.

John had grown up in Ft Pierce, working the *1715* fleet wrecks when he could. When Mel decided to look for the *Atocha* in the lower Keys, John joined on as one of the original crew. After the initial *Atocha* effort faltered, he had returned home to work the *1715* shipwrecks once again. John would become very well known by the local treasure hunting communitiy for his knowledge and success on his "home" sites.

John's personal appearance was misleading. He had a beard and a long red ponytail trailing under a perpetually worn baseball cap. He wasn't skinny or fat, and always wore long blue jeans instead of shorts. There had been some comments on our dock that he looked a little bit like a redneck Yosemite Sam. John had a fondness for Mel's own favorite drink, rum and coke. He liked to party with his crew.

Not apparent was the depth of his commitment to our trade. In addition to being articulate and technically knowledgeable about shipwreck recovery, he was driven by keen intuitive instincts and a desire to succeed. John wasn't satisfied to simply ride on the Fisher's backs, he did his own research and experimentation with new techniques and equipment, always looking for an edge. Through ignorance and ego, many of the Key West crew (myself included), initially dismissed the *Endeavor's* short term presence. After all, WE had found the amazing *Santa Margarita*, something these part time, hip boot divers could only dream about as they picked over their worked-out *1715* sites.

While the *Golden Venture* punched holes in the deep sands near where the *Jefferson* and *Arbutus* had worked, the rest of us continued pretty much as before. The *Endeavor* was employed as a mag boat for a while, and we dove a number of their hits which all turned out to be modern junk.

Because the company's emphasis had shifted more to the *Atocha*, we decided that it was time to benchmark some of the original finds on our Del Norte formatted charts. No one really knew the exact original position of the **Bank of Spain** or **Galleon Anchor** features anymore; the locations hadn't been permanently marked and were now only generally understood. I took the *Swordfish* on a digging expedition, and from the ballast, coins and old Treasure Salvors junk we found was able to finally plot the positions on the Del Norte charts. During the exercise we found the first artifacts recovered from the *Atocha* in this new year. Our luck would be quickly overshadowed.

I expected John to exclusively work the *Margarita* where the shallow sands would offer easier opportunities for the *Endeavor*. He really knew nothing of the site; rather than play catch up he decided to play his strong hand.

John took the *Endeavor* to an old spot he remembered from years ago when he worked on the *Atocha* site. There was no sand in this area so his detector expertise would really count.

The **Coral Plateau** had been ignored as long as I had worked for the company. The flat bedrock plain with its soft coral gardens had never been thought to be a big deal. I knew from old company logs that Pat Clyne had found a lone silver bar laying exposed on the hardpan years ago, but there had been no notation made as to the actual position. Ted had once told me that a rope had been stretched across the **Coral Plateau** from the **Bank of Spain** to the **Nine Cannons,** and that divers had swum metal detectors along the line without finding much. With the exception of the **Nine Cannons,** just about all the *Atocha* artifacts had been found in the deep sand.

John took the *Endeavor* to a spot on the **Coral Plateau** about 200 feet

The *Endeavor* working the **Coral Plateau**

southeast from the beginning of the **Bank of Spain's** deepening sands. Here he and his divers began free-swimming detector searches. His years of detector experience and tutelage to his young crew soon paid off.

Investigating an isolated detector hit amongst the sea fans produced a small rent in the bedrock that had concreted over. Careful hand excavation of the deepening fissure produced a broken chunk of gold bar and several large ornate pieces of gold jewelry. Two of the smaller pieces contained decayed pearls still mounted to them, and the large piece held a precious stone. Unusually, all the gold was covered with an easily removed calcareous concretion. The jewelry pieces, which looked like belt links, were obviously related.

The new find started an instant gold rush on the surrounding **Coral Plateau**. Quickly the *Dauntless*, *Virgalona*, and *Swordfish* all began their own detector searches near the *Endeavor*. John's crew went from strength to strength, some ranging further afield. They began finding scattered gold coins, often in areas that we had just searched. At the same time they kept working the area around the now expanding fissure and brought up more and more of the pearled and jeweled gold links. The "cinta" as it became known, kept growing as more of the exquisite links were found. It would become one of the hallmark pieces of the *Atocha* recovery. Without expensive navigation systems or powerful excavation equipment, the *Endeavor* had frustrated and bettered us all.

This humble pie had been served up on a hand-held metal detector loop. Our failing had its roots in the relative ease of the *Margarita's* recovery. For the first year it had been easy to find things on the *Margarita*. Even with thousands of hours using detectors on the site there had been no perceived need for our divers to experiment with the equipment or to find its strong and weak points. We accepted the detectors that the company provided at face value, assuming that each one had exactly the same sensitivity. Up until now there was nothing to suggest that we weren't the best at what we did.

Since the *1715* fleet sites had been worked hard for decades, finding things there required serious attention to detail. John and his crew really scratched for their finds, they knew how to extract the absolute maximum from their hand detector searches. Fortunately for us Key West crews, John didn't mind sharing his hard won insights. Use of these techniques spread amongst those of us who were interested and inspired continued improvements.

For a while the *Swordfish*, *Virgalona* and *Dauntless* huddled around the *Endeavor's* spot. Our divers tried their luck in the surrounding area and eventually all three boats recovered more gold coins. KT recovered an extremely rare Santa Fe de Bogotá coin, of the first made in the new world. Kane then decided to try digging in the sands near the *Golden Venture* and was soon rewarded by finding a gold and emerald ring. The *Virgalona* continued and began to find silver coins near the **Bank of Spain** where the sand wasn't too deep

John Brandon had gone with his forte, and now it was time for me to go with mine. I decided to start a large-scale circle search of the entire **Coral Plateau** from the **Bank of Spain** to the **Nine Cannons**. Using my Del Norte

Laid out on a wet suit, some of the gold still had a thin film of concretion on them. These links were the first pieces of the "cinta" belt found.

charts, I plotted the start point of each fifty-foot radius circle. We would drop buoys on those spots and cover the entire area with a careful and determined hand detector search. If the *Endeavor* had found the "cinta" in one small spot on the **Coral Plateau** what else might be found there? I hoped my unique familiarity with the Del Norte system and our improved detector skills might just pay off.

Stuart made a surprise return to the *Swordfish*. He had found captaining the *Bookmaker* frustrating at every turn, and after accidentally running it aground, decided that he had enough of the whole program. He asked to return to his old duties aboard the *Swordfish*, a request that was easy to grant. Experienced help would be valuable for the new project.

We started dropping buoys for the circle search survey at the **Bank of Spain's** sand interface and slowly worked to the southeast. Not only did our new, refined detector technique take more time, the physical nature of the **Coral Plateau** hindered quick movement. The hardpan was densely inhabited by large sponges, rubbery, purple sea fans and other soft coral colonies. Moving the detector loop in and around the obstructions was slow going. Many other animals made their homes amongst the holdfasts. Swivel-eyed groupers followed us like curious dogs, lobsters tentatively watched our passing shadows while large gray stingrays glided just over the tops of the coralline hedge.

The pace of this new procedure was much different than the *Margarita's* quickstep tempo. In the *Margarita's* shallow sands we blew holes in seconds, explored the excavation and quickly winched the boat to the next position. Our record on that site was ninety holes in one day. Here we would drop three or four buoys on positions I had pre-programmed from my chart. We would then anchor the *Swordfish* in the middle of them and dive teams would swim to the buoys and begin their searches. Even with two teams in the water at a time we rarely finished four circle searches in a day.

This wasn't just a simple survey; it was also a retort to the *Endeavor's* success. We had to show them and ourselves that we were of the same measure. Despite the tedious progress the crew remained focused. Almost immediately we started finding things.

Large lower hull spikes and a few more gold coins and cannonballs were found hiding in silted potholes or fused to the calcareous bedrock. We found that even exposed objects were difficult to see, being heavily camouflaged by sea growth. If we worked like this it would take months to survey the whole **Coral Plateau**.

Back in Key West our land base was about to change again. The property next to the half sunken *Galleon* was going to be developed into a high-rise condo and time-share. Our parking lot, docks and even the *Galleon* itself would soon be in jeopardy. By coincidence a furniture maker interested in building home décor out of weathered ship wood bought what was left of the hulk from Mel. Not long after that the Coast Guard showed up wanting whoever owned the wreck to clear it from the harbor. Mel, through luck or intuition, had sidestepped that expensive bullet just in time. Mysteriously one night the *Galleon* was burned to the waterline, and what was left had to be clam-shelled out by the developers of the property. The new development transformed Wrecker's Wharf but kept a familiar name; "The Galleon".

The romantic esthetic of 425 Caroline would soon be replaced by the echoing, bland, institutional corridors of the previously abandoned Navy Administration building in the old Truman Annex. Included in the company's new accommodations was dock space for our fleet at nearby Pier B. Before the summer season began, we completed the transition to our new headquarters and docks.

The winter in Key West had been unusually calm and mild, allowing the boats to operate nearly continuously. Soon enough, the weather upstate was good enough for the *Endeavor* to head north, back to the *1715 Fleet* sites. During their few months prospecting the **Coral Plateau** they had located gold coins, a small piece of gold chain, a gold bar fragment, silver coins, musket balls and 20 ornate links of the "cinta" belt. Treasure wasn't the only thing John Brandon found during his Key West stay. When the *Endeavor* departed for Ft. Pierce our cook Debbie was aboard, enroute to a new life with John.

Denny Breese and the *Tern* also left. With the *Golden Venture* flailing around in the sands to little effect, it was decided to put Dick Klaudt aboard to help guide their progress. Dick had learned well while doing fine-edged surveys aboard the *Tern*. He knew the Del Norte system and was good at detailed position plotting. Dick tightened up the ship's excavation patterns and was stubbornly insistent on thoroughly working along the theorized scatter trail. His efforts would soon be rewarded.

The company's finances were becoming necrotic again. The full-on pace maintained by the Treasure Salvors fleet during the mild winter had gobbled up funds, money that would have been normally expended during the summer dive season. Paychecks were becoming more irregular; sometimes several weeks would pass with no money for payroll. We were stretched to put divers on board the *Golden Venture* and there wasn't money to hire more.

To avoid returning to port where there was no waiting paycheck, several divers including KT decided to remain at sea aboard the *Golden Venture* while their assigned boats returned to Key West for reprovisioning. She would spend three months without setting foot on soil, alternating between working aboard the *Swordfish* and the *Golden Venture*.

Since there was always a boat on site now and with no extra crew to spare, Kane towed the now abandoned *Arbutus* north in the shallows between the sites and anchored it well out of the way during the summer. Like the departed *Galleon* in Key West, it shortly succumbed to age and neglect, quietly sinking with its deckhouses and mast protruding from the water. Freshman Treasure Salvors divers cheered the end of forced *Arbutus* duty much the same as my contemporaries had celebrated the demise of *Galleon* patching.

The long spell of good winter weather had strained more than the company's accounts. Everyone was already getting burned out and testy, and the heat-induced fatigue of summer hadn't even begun yet. This only amplified KT's continued issues with some of Treasure Salvors' divers over her presence amongst them. Our personal situation had begun to unravel as well. My own inability to invest more of myself in our relationship and effectively squelch her antagonists added another layer of abrasion.

Up to this time the *Swordfish* was the only recovery boat that continuously updated its own site maps, plotting our own excavations as well as everyone else's. Both the *Dauntless* and *Virgalona* logged their own activity coordinates, but turned the information in at the end of each trip for Ed Little's master chart compilation. I was surprised when Kane's First Mate Andy Matrocci approached me about learning more about plotting and Del Norte navigation. I taught him the basics, and he was soon updating charts for the *Dauntless*. They began to use the system as I had, not simply to record their excavations but as an important tool for site interpretation.

During a visit to Key West, Ted informed me that Mel had offered

After the *Arbutus* sank, it still found use as a platform for a Del Norte tower.

the *Swordfish* as a temporary replacement for an ailing tugboat towing a huge floating dry-dock far out into the Gulf Stream. It would be impossible to bring the dry-dock into Key West's constrained harbor, but the tug needed to come into Key West for emergency repairs, and if the *Swordfish* assumed the tow for about 30 hours, Treasure Salvors would be paid a lot of desperately needed money. Until now, I had no idea that the company's finances were anywhere near this bad. Almost as scary was the news that the mailboxes and their framing would have to be cut off, as they would be in the way of the towing bridle. The thought of removing our prime excavation tool was like having an arm cut off. Mel was obviously desperate.

With the rest of my crew temporarily placed on other boats, Stuart and I completed the surrogate towing job. As promised, the towing company made good on its financial obligation adding some black ink to our ledger. The *Swordfish* quickly returned to the *Atocha* site with its blowers reattached, but the quick infusion of cash didn't last long. Why was money becoming such a problem again?

The *Swordfish* continued with its lonely **Coral Plateau** circle search program toward the **Nine Cannons** while the other boats churned through the sands. I was inwardly pleased with the new level of skill that my crew displayed during the long detector searches. We continued to carefully locate and plot the scatter of artifacts as we progressed, though as yet there was still nothing found as showy as the "cinta". KT's searching dexterity really surprised me; in the slow and detailed work her detector abilities had proven equal to the best of my most experienced crew. If there was something significant where we were looking, I was certain we would find it.

Almost two months of circle searches finally terminated at the bottom's drop off into the mud at the **Nine Cannons**. My investment of combining the precision of the Del Norte and an exacting detector survey had ended in an unsatisfying way. We had found more artifacts than originally anticipated, but it would have been nice to be able to include some unique or precious items to the list.

As I reviewed my chart one last time I noticed two small areas along the middle of the plateau that had been missed in the geometry of the already completed circle searches. The missed areas were so small that I was tempted to ignore them and get on with an entirely new project; we were all pretty tired of weaving the detectors around sponges and gorgonians. After waffling, I finally decided that since we had gone to such an extreme amount of trouble to do a thorough job we couldn't leave without finishing it entirely. I plotted the locations of the spots and we dropped a buoy on each one. Since the areas to search were so small, Stuart and Michael Moore would dive them both.

Minutes after beginning the search, both were back on the surface laughing and asking for a rope. They had found a silver bar, heavily encrusted with marine life, lying exposed on the bedrock. After plotting the bar's position and recovering it, they quickly finished the search of the first area and moved over to the second.

It was almost a reprise. This time it was a large gold bar, also somewhat encrusted, residing exposed amongst the sea fans. There were big smiles all around as the now completed project had rewarded our efforts with a special bonus. So far there had been only five silver bars found on the *Atocha*, and those had been located years ago. There were twelve hundred bars listed on the ship's manifest. At this rate of recovery, none of us would live long enough to find them all.

Over the last year we had heard of a particularly irritating trend initiated by certain shore bound employees. Media types would often visit the office to talk about new developments in the search. Often the company "representative" actually implied that he himself had personally found recent notable artifacts.

Half seriously Stuart and Michael posed for pictures with their still wet *Atocha* bullion, displaying handwritten signs denying the discoveries were made by the perp. The gesture would mean nothing in Key West, but it made us feel better. There was little time to dwell on our eleventh hour suc-

The **Coral Plateau** gold bar

cess; things were starting to happen under the *Golden Venture's* stern. We too moved into the **Quicksands**.

Dick Klaudt's stubborn, incremental excavation through the **Quicksands** had begun to pay off. Gold bars, found in pairs were being recovered as the *Golden Venture* continued to dig to the northwest beyond the **Galleon Anchor**. The bars themselves were very unusual, being nearly identical and having no tax stamps or other markings other than carat markings. They were all of the same purity, twenty-three and three quarters carats. Any reservations Ian Koblick may have had about Dick's methodology were now gone.

All gold bullion found so far on the *Atocha* or *Margarita* had varied in size, shape and carat. Legal bars or discs were covered with tax stamps showing that the government had taken their due. Unmarked smuggled bars had been found too, but they tended to be small, unmarked and of lesser purity. There was nothing in our inventory of prior recoveries like these.

As more of the bars were found, their uniformity and extreme purity suggested something other than smuggled bullion. Since the Catholic Church wouldn't have had to pay tax on their property, maybe the Church had owned the very fine gold. We needed more clues.

The summer ocean may have been calm, but aboard the recovery boats there was an undercurrent of tension. The veneer of graciousness barely covered the competitive jealousies between the boat crews. Contractors and their boats were often viewed as second-class outsiders anyway, and if a non-company boat did well it often heightened the stress.

Sometimes those individuals who found unique or flashy artifacts got just plain weird, becoming egocentric prima donnas. One diver involved with a significant find got so annoying that he would eventually be kicked off the ship on which he was worked and was permanently blackballed from all the others.

Despite the good chow, nice accommodations and improved chances of finding artifacts, it was still hard to get T.S. divers to volunteer for work aboard the *Golden Venture*. Days aboard were exceptionally long and often the same divers dove all day everyday from sunrise to sunset with no rotation.

As Dick guided the *Golden Venture's* digging pattern, it became clear

The *Golden Venture* excavating in the *Atocha* **Quicksands**

Sunburned after months at sea, a tired KT displays two gold bars she found under the *Golden Venture*.

that the gold bars were being found along a thin straight line projecting further to the northwest. This became known as "**the gold line**" and all the other recovery boats jumped in to work on the theoretical extension of this track. Since Kane's *Dauntless* was digging just southeast of the *Golden Venture* I tried my luck just at the beginning of the sand interface near the **Bank of Spain**.

I had "borrowed" KT back from the *Golden Venture* to help fill out my own crew. While diving under the *Swordfish's* blowers, she found a unique artifact. It was a small gold clasp in the shape of two rampant lions; probably a fastener for an article of clothing such as a cape. When Mel called on the radio later that day I described in detail the interesting find.

Mel had always been known for his chronic exaggeration but this time he really outdid himself. By the time the *Swordfish* arrived at the company's dock several days later, he had convinced a number of people including a bewildered reporter that we had found life-size, solid gold lions. Imagine the scene when I produced the inch and a half long clasp to the anxious crowd waiting for us at the pier.

Digging the deep sands on the *Atocha* was like moving in slo-mo. Each excavation took from twenty to thirty minutes to expose the bedrock. Checking each hole took even longer. Since the sands continually shifted with storms and the tide, artifacts had worked their way down to the bedrock. The sands sucked down even our anchors; we never had to worry about them dragging like on the *Margarita*.

The large flaked, calcareous sand had the friendly look and texture of uncooked oatmeal and the deep craters created by our digging gave the divers a haven from the current. It was fun to glide down the berm onto the exposed bedrock and the sand's light, reflective color considerably brightened the underwater vista. We paced ourselves, trying to dig at least ten to fifteen holes a day.

KT had returned to work aboard the *Golden Venture* and would find more of the strangely uniform gold bars. Late one July afternoon she was diving with Rick Ingerson in the latest excavation. He was using a large search loop detector that was good for finding deeply buried artifacts, but difficult to pinpoint the hits with. KT had a small loop detector that operated just the opposite.

Rick got a detector hit in the unstable wall of the berm but was having trouble trying to locate the source of the hit. KT used her detector to indicate the correct position as they worked together to try and hand dig out the cascading slope. Further excavation by the *Golden Venture* failed to uncover what caused the hit, so in desperation they removed their fins and tried to use them as a cofferdam while frantically hand digging. Over and over each rechecked the spot with their detectors to confirm that the elusive hit was still there.

The day was getting quite late and their air was running out. While KT struggled to hold back the avalanche of sand, Rick worked his arm deep into the berm and felt something. He pulled the item out of the sand bank, looked at it and handed it to KT. It was in the shape of something that was regularly found in the deep sand. Lobster fishermen baited their traps with punctured sardine tins. Rick assumed the item was simply more of this common trash and dismissed it. KT wasn't so sure - something inside rattled. They both headed for the surface.

An advancing evening storm altered the sunset's colors and the wind was picking up. Dick Klaudt was waiting on the back deck as the divers came up the dive ladder. KT brought the item up to the surface and showed it to Dick. It still rattled.

Huddling in the lee of the ship's gunwale, Dick and KT studied the nature of the box. The encrustation around the edge easily flaked away, and after treating the concretion with some diluted phosphoric acid, with little effort they were able to open the top. For a moment in the dramatic light all they could see the bottom of the box was green. The shock forced an automatic response to quickly close it. After a quick breath they opened it again.

The box contained an ornate gold cross set with seven large emeralds. Also in the box was a large gold man's ring also boasting a huge emerald. It was as unexpected as the cinta had been and just as dramatic. There was no denying the religious nature of these items. The excavation from which it had come was right on the gold line. This certainly added to the theory about the untaxed gold bars belonging to the church.

The next morning after the *Swordfish* repositioned some of the *Golden Venture's* anchors, I visited with Dick Klaudt to update my charts with his latest data. After seeing the gold cross and ring, we had to agree that the *Golden Venture* had lived up to its name.

Treasure Salvors' company boats spent the rest of the summer work-

The emerald cross and ring on the silver box they were found in. Blue enameling decorated the gold of the cross.

ing the sands all around the *Golden Venture's* advancing position. There were more religious artifacts found just west and northwest of the emerald cross-area, but the trail rapidly cooled. I tried digging along the southeast extension of the gold line near the **Galleon Anchor.** At first all we uncovered was old T.S. junk, discarded years ago. Ironically, one diver actually found a twenty-one carat gold bar when he rolled over an old compressor laying on the bedrock.

1982 had been a very long season and the late summer's heat left us all withered. The divers like KT who had jumped from boat to boat without going ashore for a break were really fried. Our first time together while not working was in September when we both drove to Wisconsin to take care of personal business. The trip did not go well.

I had never been involved for this long or this closely with anyone before, and somehow I had lost my way. Aboard ship I acted definitively on everything the *Swordfish* crew did, but my relationship with KT suffered from my own lack of conviction. Despite the long hours of travel together we arrived back in Key West with our relationship even more fractured.

KT quickly rejoined the *Golden Venture* back out on the site. At least it was a reprieve from our own unsettled state. Tom Ford, who was working aboard, recognized her abilities. He selected her and another diver for a high detail re-survey of the "cinta" area using a new experimental detector. They would work out of a small boat all day, returning back to the *Golden Venture* at night. The project took several weeks but paid off when KT found a gold clasp belonging to the cinta.

As fall arrived the *Golden Venture* suspended operations on the *Atocha* and returned to Key West. Ian Koblick had other commitments for the ship; it was time for them to go. It had been a hard season to qualify. For all the valuable and dramatic artifacts found over the year, our knowledge of the site was not much greater than before. All recoveries had been made in areas found years ago and were more an indication of improved technique and equipment than anything else. The departure of the *Golden Venture* left us with little to consider, except the company's spill of red ink.

There was no resolution between KT and me either. We were no longer talking, just arguing. I was afraid to commit more to the relationship. In an act of frustration and hurt, she packed up her things and left for Wisconsin to regroup.

I was deeply tired and depressed about the bad ending to the romance and the company's cancerous financial issues. Focusing more on work gave some diversion. While reviewing the charts of our **Coral Plateau** survey I noticed a detail I had missed before. The artifacts found between the **Nine Cannons** and the **Bank of Spain** were not in a straight line, but along a subtle though definite arc. The precision of the Del Norte mapping system really defined the trend.

The arc could only have been created as the *Atocha's* wreckage was pushed by wind and waves that had changed direction as the storm itself moved. That implied that the artifact track might not have originated to the southeast near the **Patch Reef** as commonly thought. Had this "clocking" wind started more from the east?

I talked to Ted, Bob Moran, Stuart and anyone else who seemed interested about my theory. They mostly agreed that there might be some merit to the concept. I knew the really hard sell would be Mel, who had always believed strongly that the *Atocha* had originally struck the **Patch Reef**. I went to Mel's office armed with charts to plead my case.

Mel listened passively as I told him of the significance of the arc. The presentation seemed to be going well when all of a sudden he leaned forward and started pointing all around the site map. "Well, I think it hit here on the **Patch Reef** and the storm carried it to the north where it grounded here. Then the current moved it back south and it stuck here. Then the wind blew from this direction and took it..."

I should have guessed that simple explanations weren't dramatic enough for Mel's active imagination. He had dismissed my idea and spontaneously improvised what would later be called "the pin ball theory". Without arguing my case any further I rolled up my chart and left.

KT wasn't the only employee to depart. Health issues had affected Momo, and he had gone back home for good. The *Virgalona* was parked since there really wasn't money to operate it anyway. Some of her crew were absorbed on other boats, others left completely. As paychecks grew more and more rare, disgusted and burnt out divers from all the boats sought out new employment. The office staff dwindled as well. Stuart had reached his limit, and joined the growing number of ex-employees.

After the money free-for-all of a year ago it was hard for the boat crews to understand why things had gotten so bad. One big reason was unresolved legal issues over the *Margarita*, both with Bobby Jordan and the State of Florida. As yet there was no final decision regarding the state's ownership challenge for the wreck, and Bobby's case was still awaiting a final ruling. Until these two problems were worked out, Treasure Salvors couldn't give its investor's returns on their money or sell any of the treasure. New or returning investors had become very leery until the wreck's final ownership had been legally decided.

As the company slowly collapsed we tried to continue operations. Both the *Dauntless* and the *Swordfish* made trips, usually short crewed and often just diving on mag hits from the *Plus Ultra*. Nothing new was being found and the summer's remarkable discoveries now seemed a very long time ago.

I found out years later that Stuart and his old friend Larry Beckman had discretely joined forces to find the *Atocha* themselves. They were interested in a spot on the outer reef south-southwest of the Marquesas, outside of Treasure Salvors' protected area. Knowing how Mel would likely react if he found out, they carefully kept their distance from any Treasure Salvors' boats, operating their search gear from Larry's small fiberglass runabout. The *Atocha's* main pile stubbornly eluded them as well, and they eventually gave up.

With little good news about the company's prospects, I was getting pretty nervous about my own future. Like the other active Salvors captains I had never needed a license to run the *Swordfish* since we carried no paying passengers. I had logged enough sea time to qualify for one, but there was a rather long test of practical knowledge for which I had to study. It was time to act now - if I got the license, I might be more employable somewhere else. At least I'd get something out of the whole experience.

Arriving under the radar was Alice Green. She turned up during the Thanksgiving break looking for a playmate. Through the rumor mill she had already found out that I was no longer "attached," and offered to spend time with me while she was in town. Her uncomplicated nature had a calming effect, but I wasn't great company. I was just too stressed about my future at Salvors, studying, and the unhappy ending with KT.

Within a month I had my license; I was a real Key West captain with the papers to prove it. Since just about all my crew had left or been laid off, the accomplishment felt hollow. I still clung to the hope that things would change, that Mel would somehow find money and everything would be OK again.

Mel and Deo tried to put a positive spin on things by once again having the traditional company Christmas party at their home. Much of the quiet talk amongst past and present employees there was about Treasure Salvors' fragility. I personally caused some gossip arriving with Alice instead of KT. Our brief reunion would end a few days later when she returned to school.

The New Year of 1983 at least brought some change. Mel had obtained an old navy warehouse on the Truman Annex property and was going to convert the multi-floored, empty building into offices and a museum. If the museum could be completed before the company's demise there would be some income from admissions.

KT returned from Wisconsin, determined now to make a life for herself within Treasure Salvors without me. She, Dick Klaudt and a few others were employed converting the warehouse into exhibit space. Tom Ford had assumed command of the *Bookmaker* and on his rare trips to the site KT served as crew. She and I acknowledged each other but there was still too much bitterness between us for much else.

Bobby Jordan's case was finally resolved and the State of Florida also conceded that the *Margarita* was out of their jurisdiction. The company now held unchallenged title to the site. The finds could now be divided with the investors and hopefully that would inspire further investment. The company could now sell some of the treasure as well. Mel and Deo scraped together money for an impromptu but joyous party at the Pier House. At last we had

The new T.S. headquarters was an ex- navy warehouse.

something to celebrate.

Perhaps the feel-good atmosphere inspired me. I approached KT at the end of the party in an awkward attempt at some sort of reconciliation. She was very suspicious, still bruised by our parting. If we were going to function as anything but co-workers it was going to take a lot of work.

The expected financial turnaround didn't happen. What money did come in was immediately blotted up by angry creditors. T.T. Hargraves had once adventured as a mag operator during the early search for the *Atocha*. Now instead of marking anomalies she and her daughter Karen had the unenviable task of crisis bookkeeping. Their tiny office was a tilting repository of past due notices.

Operations completely stopped. A strange policy had evolved; those employees who quit or got laid off were immediately paid all monies owed to them. The rest of us still on the books accumulated IOU's. Every other week we might get thirty or fifty dollars, listed on the check as an "advance". It was getting really hard to hang on.

For a while, KT's attention was held by a study she had undertaken at Duncan Mathewson's suggestion. She researched the origins of the markings on all the gold bullion that had been recovered on the *Atocha* and *Margarita*. It was the first real attempt to catalogue and interpret the 1622 fleet bullion. How ironic to be surrounded by gold all day while the company was foundering.

The office staff was just about reduced to the state that existed when I first showed up at the *Galleon*. Duncan, Ed Little, Bleth, Don Kincaid, Pat Clyne, and the conservatorial staff were all gone to other opportunities. John Green had started working privately for Bob Moran. Kane rarely came around, his crew was gone too. Everyone else in the operation had been laid off now except Tom Ford, KT and I. Our sole function was to try and keep the boats afloat and their systems functional. If something was broken there was no money to fix it. The whole situation was very depressing.

If the company was in an unrecoverable tailspin, I was determined not to ride it into the ground. My best parachute was Pam Raabe, Stuart's Australian girlfriend. A year ago in return for the hospitality I had shown her, she had offered accommodation and perhaps a job contact if I ever wanted to visit Australia. When things at Salvors started getting really bad I had written her. She had quickly replied that the offer still held. All I had to do was figure out how to get there.

Little by little KT and I began untying the tangle of bad feelings between us. Gradually the anger and stubbornness began to subside, even if our self-absorbed world at Treasure Salvors was circling the drain. At least this one thing appeared to be improving, even if it took a lot of work.

Leah informed us that the company was trying to organize a division of the *Margarita* finds for its investors. As employees, we had always been verbally promised a tiny portion of the division too. How much or what had never been clearly defined.

When Mel's mood had been brightened by a long rum-and-coke

lunch, he sometimes pronounced we would all be millionaires. Old time employees hinted that we would be lucky to get a few coins. None of us really expected to get much, but if there was ever a time for a change of fortune, this was it!

A few days later I was presented with a list of things that I would receive as my part of the division of the *Margarita* artifacts. Amongst the items listed were a length of gold chain, a gold coin, a number of silver coins and even a restored dagger. The problem was that it might be months before any of the items could be made available to the investors or employees. The dramatically reduced office staff was slow to the task.

I had no idea of the worth of the things I was supposed to get. Ted's only reply to my query was, "Whatever someone will pay for it." After several months of no real income I needed money now, especially if I was going to make it to Australia.

Bob Moran offered to buy the rights to my division sight unseen if the price was right. I was at a loss as to the value. I naively figured that as soon as all this treasure made it to the market that the price would go down to nothing. With that mindset, I found it too tempting to resist Bob's offer. I sold my share to Bob for little more than the amount it would cost me to get to Australia.

All indications were that Salvors would be lucky to survive another month. It was time to act. If I didn't quit soon the company might never have the money to catch up on the months of delinquent back pay. I bought my plane ticket to Sydney, and with real reluctance gave Salvors three weeks notice. That gave the bookkeepers some time to try and come up with the money the company owed me.

The reconciliation between KT and me was now gaining ground. We decided to both go to Australia. She wouldn't be going with me; she wanted to get her captain's license before she quit Salvors. The plan was for her join me there in about a month. We even tiptoed around the subject of marriage.

Weeks later, in April, I was on a plane over the Pacific. My time with Treasure Salvors had been a defining but incomplete event, much like taking in a breath and not exhaling. By now the company was long gone like the *Atocha* and *Margarita* themselves.

# Re Do

One year later, in January of 1984, KT and I arrived back in the states. We were married. Our return trip from Australia ended in Wisconsin for no other reason than that was where our possessions and family were. Neither of us had a clear idea what we would do next.

Within twenty-four hours of our arrival, Deo Fisher telephoned my parent's house looking for us. Apparently the last rites given to Treasure Salvors had been premature. Against all probability, the company had been resuscitated. It was becoming viable again and they wanted us back.

Deo explained that Mel wanted to expand work on the *1715 Fleet* sites. He wanted KT and I to run a detailed magnetometer survey on all the established wrecks there as well as explore for some of the missing ones. We would have a month or so before the project began to get organized.

This was fantastic news. With this unexpected phone call we were back in business. Our preference would have been to go back to the *1622* sites, but at least we were again getting another chance to work Spanish shipwrecks. It would be a great education examining the famous sites where the great Florida shipwreck "gold rush" had started firsthand. The discovery of these wrecks had been the genesis for the industry of shipwreck "treasure hunting", and through its growing pains diverse perspectives on marine archaeology evolved. Many of the original participants such as Kip Wagner, Mel and Deo, Fay Feild and others had become legendary.

It was hard to believe that a whole year had gone by so quickly. Australia and its people had drawn us in. We were able to employ many of our hydrographic skills there and during leisure times explored some of the limitless possibilities at hand. With our personal issues now under control we took the time to get married on a friend's boat in Sydney harbor. It would have been easy to stay there forever but a persistent, low gain intuition drew us back to America.

We spent a few weeks rebuilding an old Land Rover that some friends had given us as a wedding present and made our way back to Key West. After a happy reunion, Ted and Leah accompanied us in their motor home to Ft. Pierce where we would be based for the season. Leah helped find an apartment for us and Ted helped make sure our survey boat and its gear were ready to go. When the preparations were finished they returned to the Keys and we got to work.

Mel had won title to the *1715 Fleet* sites after the state of Florida's battle with him over ownership of the *Atocha*. After years of legal combat over the rights to the *Atocha's* riches, the state had lost its final case against Mel. The judge was going to force the state to reimburse Mel for the money he had spent defending Treasure Salvors, but Mel made a different proposal. Instead of paying Mel, all the state had to do was grant him ownership of the still valuable *1715* wrecks.

It was a good deal for everyone. Since the wrecks were legally in state waters, the Florida government got a twenty-five percent choice of all artifacts found. Instead of having dozens of different interests competing for areas on the wreck sites, Treasure Salvors now selected and oversaw those that were viable. The *Endeavor* was the only boat owned by Salvors that regularly worked the *1715 Fleet* sites, but the company got a percentage of everything found by these subcontractor salvors as well.

KT and I quickly adapted to the small-scale nature of operations on these close-to-shore wrecks. Our survey boat was only a twenty-two foot Mako outboard with a magnetometer and basic equipment for the job. Every night we returned to our little apartment. We spent the summer working on specific wrecks as well as looking for new ones. The local names, *Colored Beach, Sandy Point, Rio Mar, Corrigan's* and *Cabin Wreck* gradually became familiar to us as well as each site's unique properties. Though the Spanish names of the lost ships were all known, nothing definitive had been found to absolutely identify which site was which.

There were other wreck sites as well. *Green Cabin* was a shipwreck whose artifacts dated older than the *Atocha*. The *Spring of Whitby* was an English wreck whose name was discovered upon the recovery of the ship's monogrammed bell. There were World War II wrecks and Civil War wrecks, in fact we located a wreck about every quarter mile along the twenty-five mile corridor of our coastal survey. Most of the debris and artifacts from them was concentrated in less than ten feet of water.

It was surprising how different the diving conditions were from the Keys. Here the water was cool, requiring a full wet suit for extended diving, even in summer. This close to shore the waves constantly churned up the bottom, the water was often a green-gray liquid grit. Crumbly near-shore reefs of Coquina bedrock resembled a lumpy, fragmented terrazzo.

During the course of the summer we got a couple of unexpected phone calls. Barry Clifford phoned and wanted to know if we would be interested in helping him look for the pirate vessel *Whydah* off Massachusetts. KT and I were flattered but reluctant to leave Mel's employ. I gave him Stuart's phone number as a possible alternative.

The other call was less appealing. A contractor who had once worked the *Atocha* site wanted to hire us to run a hundred and forty foot recovery vessel on a very illegal sounding, grab-and-go shipwreck project in the eastern Caribbean. The name of the recovery boat mentioned sounded something like *Samba Rock*. He got quite angry when we turned down his offer.

During our time on the *1715* project we were lucky to work with some of the most successful operations. John Brandon and his crew collaborated with us on several sites and we enjoyed the brief opportunity of working with Bob and Margaret Weller off their own recovery boat. By the end of the summer we had generated lots of hits, data and site maps that helped advance the overall salvage effort. When fall's strong winds closed the season, KT and I headed back to Key West to rejoin the *1622 Fleet* search.

On our return to the Keys, we learned of some notable developments that had occurred during the summer. Bob Moran had put the *Plus Ultra* to work again and had continued surveying, this time some distance to the northwest of the *Golden Venture's* last finds. One calm day, continuous magnetometer problems so frustrated Fay Feild that he jumped into the water to cool off. Imagine his surprise when he looked down and saw a bronze cannon lying exposed eighteen feet beneath him.

This new find, appropriately named **Fay's Cannon,** triggered an intensive survey and search effort further to the north and northwest. Sure enough, about a mile further north was the next find. Two large galleon anchors, obviously from the *Atocha*, were found imbedded in a sandbank. As was so often the case, these new finds generated more questions than answers. Even with all the concentrated work between these **North Anchors** and **Fay's Cannon,** nothing else had been found.

Kane had begun prospecting on the other end of the *Atocha* site. He had taken *Dauntless* back into the deeper water mud area not far from the **Nine Cannons.** Patient searching had rewarded him with a wispy trail of barrel hoop fragments and spikes that seemed to lead to the east-southeast. There were large distances between each artifact, which made detecting their presence amongst all the modern seafloor junk extremely difficult. As *Dauntless* slowly advanced farther along it became obvious that this almost invisible track didn't lead to the **Patch Reef** as per Mel's favorite theory. This new direction, nicknamed **Kane's Trail** had been strongly implied by the results of the *Swordfish's* **Coral Plateau** survey a year earlier.

Ex-employee Donnie Jonas had converted an older oilrig supply boat called the *J.B. Magruder* for treasure hunting. He had contracted with Mel to use it on the *Atocha* site during the time KT and I had been in Australia. Years earlier as a teenaged high school dropout, Donnie been hired to crew aboard the *Northwind* and was acting as ersatz engineer the night it rolled over and sank. He was lucky and resourceful enough to extricate himself from the dark, upside down flooded engine room just before the doomed boat went down.

Donnie was a good friend of Kane, and had periodically worked for Treasure Salvors over the years. Despite having little formal education, he used his high intelligence and local connections to prosper in Key West's alternative economy. Donnie remained intrigued by the *Atocha* and used some of his growing fortune to buy and outfit the *Magruder*. When Donnie and Mel couldn't come to terms on a renewal contract, Donnie angrily parked the boat at our dock, refusing to take it out again.

The *J.B. Magruder* had been well equipped for its purpose. It was nearly one hundred feet long, smaller than the clumsy, overly large ships so many subcontractors used, but big and stable enough for comfortable extended work. It was air-conditioned, had expansive living areas, powerful winches, a broad back deck and large sturdy "mailboxes". I eyed it jealously, imagining how perfect it would be out on the sites. Too bad it was idle, it was exactly the kind of boat Mel needed.

The most difficult news for me to accept was that Tom Ford was now captain of the *Swordfish*. He survived Treasure Salvors' fiscal relapse and had been appropriately rewarded for his efforts while we had been in Australia. I respected Tom and although he deserved the honor, it was hard for me to think of the *Swordfish* as someone else's.

With artifacts recently being found on either end of the site, debates were ongoing between the employees as to where the *Atocha's* main section lay. I personally felt that the items Kane had found in the mud were exciting, but there were those such as Ted who were convinced that the whole shipwreck was likely to be up in the shallow **Quicksands**.

The company's base and offices were up and running in the converted navy warehouse. The Mel Fisher Maritime Museum Society, a non-profit offshoot that operated the formal museum display area, now occupied one half of the building. Treasure Salvors had donated to the Society a large collection of artifacts, including many of the priceless or unique items found on the

Tom Ford replaced me as Captain of the *Swordfish*.

1622 sites. Salvors itself resided in the building's other half, and if visitors were so inclined they could buy coins or inquire about investment opportunities there. The two entities ran independently but had a close symbiotic relationship.

With no better options open to me, I accepted the position of captain of the *Bookmaker*, and was directed to continue with more survey work on the *Atocha* site. Dick Klaudt was assigned to the boat as well, and together we would operate sidescan sonar and magnetometer surveys along with the *Plus Ultra*. Dick and I worked very well together but I found running the *Bookmaker* rather one-dimensional after the *Swordfish*.

Our activities didn't include any diving or excavating, just ranging back and forth across invisible electronic grid lines while racking up anomalies. The lightweight *Bookmaker* was a cork in any seas, and the aerodynamics of its superstructure drew diesel exhaust fumes directly up into the flying bridge where the helm was located. As Stuart had previously found, it always seemed to need mechanical attention. I enjoyed Dick's company but I found my time aboard uninspired.

KT made the occasional trip with us but she was more often involved in curatorial work in port. Ted and Leah left on a long planned, extended trip to Mexico in their motor home, so I inherited Ted's job as logistics manager and KT assumed all of Leah's curatorial duties for the months they would be gone. We even got to house sit 417 Cactus Drive for them. It was posh living, keeping us shore bound for the duration of their trip.

Even though the company appeared to be thriving again, not everyone was happy. Bob Moran and his wife Marisha had been growing more and more critical of Mel's business practices. From Bob's perspective, too much of the investor's money was frittered away on ill-conceived whims or family issues; invasive nepotism often rewarded inappropriate behavior.

Mel had always run Treasure Salvors as if it was a small, family business. In reality it had dozens of employees, a board of directors and hundreds of stockholders. Bob Moran himself held a fair amount of company stock. He and Marisha gradually began lobbying other stockholders about their concerns, trying to gain support for change.

A coup-de-tat was in the wind. Bob wanted to end Mel's direct access to the company funds, and take him out of the real decision making on the *Atocha* expedition. Problematic family members would be terminated. Bob wanted Mel to function soley as a money raiser and front man for Treasure Salvors. Bob and a new board of directors would assume real control over the company's destiny.

With a stockholder's meeting rapidly approaching, both camps began vigorous campaigning for the stockholder's votes. Mel's battle cry was, "They're tryin' to steal my company!" A number of the office staff did their part to make sure the stockholders would remain loyal to Mel. Perpetual Fisher family advocate Pat Clyne jumped in like a pit bull.

Bob Moran also had his supporters. As the all-important meeting approached, the rhetoric and panic on both sides escalated. It was hard to know who the eventual winner would be.

The boat crews largely took a wait-and-see attitude. Bob had raised some legitimate concerns, and some crewmembers had complained of an implied caste system between family and non-family employees. From the beginning of my employment, I had assumed that nepotism was the price paid for working in a family-run business. Other crewmembers were less resigned.

In the end, Bob's crusade failed. Mel proved to be more popular with the stockholders; it was his unvarnished charm that had drawn them to invest in the first place. Most were strongly sympathetic about the *Northwind* tragedy and Mel's "David and Goliath" battles against the government. Bob was less well known to them and may have come off more as a disgruntled employee than a corporate savior. Shortly after the vote, Bob and Marisha were terminated from the company, leaving Key West for a horse farm up north.

Though the *Plus Ultra* would no longer divine the waters off the Marquesas with its electronic magnifying glass, Bob's contribution to the *Atocha* search did not fade with his dismissal. He and his crews had cataloged and plotted many thousands of anomalies, most still waiting exploration by the recovery boats. These plotted "hits" looked as numerous and random on the charts as stars of the Milky Way. Was one of those stars a lucky one?

Kim Fisher permanently returned to Key West. For years he and his wife Jo had lived in Michigan, raising a growing family of their own while he attended college and then law school. Only during school breaks or holidays would they return to the Keys for a visit. It seemed they had intentionally isolated themselves from the project after the drownings on the *Northwind* and the accidental death of Nikki Littlehales while Kim captained the *Southwind*.

Kim's personal tide began to change at the conclusion of his studies. He was drawn to be back with his parents and siblings, to live again in the alluring drama of the search. The move south not only ended years of self-imposed isolation from the *Atocha*, it also proved fatal to his marriage.

Mel was thrilled with his son's return, promoting him to captain of the *Bookmaker*. Key West itself offered Kim many persuasive temptations with little consequence, and after years of stable Midwest living, he lost his internal compass. When their relationship dissolved, Jo and the kids went back to Michigan while Kim foundered.

Luckily, Kim had reserves, which no one knew about. Like his sister Taffi, he had a hard nugget of inner strength deep inside that allowed him to gradually sort out his life and regroup all the stronger for it. If he could not repair his old life, Kim would build a new one. During a brief exploratory expedition outside the country, he met an attractive and pragmatic young woman named Lee who eventually would change his life.

Lost and Found

    KT and I had really enjoyed the responsibility of our temporary positions. When Ted and Leah returned from Mexico, we were very happy to see them but not quite sure now what our next jobs within the company might be. There were no current promising developments on the *1622* sites and with no pressing responsibilities we decided to take a vacation and tour the British Isles.

    Six weeks later, in late May of 1985, we returned. Mel and Ted were waiting with a new assignment. A new subcontractor with a large boat had started working the *Atocha* site. It had originally started digging in the **Quicksands** but had not found much. The decision was made to shift digging operations to the mud zone and continue excavating along the proximity of **Kane's Trail**. After weeks of finding nothing, a sudden and isolated discovery of some gold bullion had gotten everyone's attention.

    After the discovery, somehow the scent of the trail had become lost and the ship had returned to digging only empty holes. Mel felt that to become productive again, the subcontractor's captain needed assistance. KT and I were sent out to observe the operation, and give advice on the excavation plan and dive the excavations.

    The *Saba Rock* was a weathered, ex-oil rig supply vessel about one hundred and fifty feet long. The bridge and living accommodations were all combined in one structure near the bow. Two thirds of the ship's length was taken up by an open back deck void of anything except a centrally located diesel powered winch system. The boat had been fitted with the largest mailboxes that we had ever seen, hinting at the power of her massive engines.

KT and I were not the only Treasure Salvors divers who worked aboard. We replaced a previous team, and later would be joined by others during our extended stay. There was a windowless, air-conditioned portable living unit similar to that of a house trailer located at the forward end of the back deck near the deckhouse. This had been set up for Salvors' personnel. The rest of the ship's inhabitants lived in the deckhouse.

The ship's regular crew was a unique dichotomy. All the maintenance and grunt work was taken care of by a number of black Dominicans. None spoke English, and due to their legal status, were restricted solely to the *Saba Rock*. Viewed as illegal aliens by U.S. customs, they could never take the opportunity to ride into Key West aboard one of Salvors' boats. The Dominicans labored all day every day as if the world didn't exist beyond the *Saba Rock's* hot steel ramparts.

Saba Rock's captain and his girlfriend were the other half of the unlikely mix. Ed had been a high school football hero; blonde, curly haired, handsome and still powerfully built, though somewhat softened by beer exposure. Susan had been a cheerleader and retained the looks and personality of one. Neither one of them had ever done anything like this before.

Susan was quite proud of Ed. She was careful to tell us that this little adventure was just a brief sabbatical from their real plans. This fall Ed would be going back to college to work toward a law degree. They would be getting married once his studies were completed; a prosperous and happy future was assured.

Captain Ed mouthed the same sentiments, but without enthusiasm. He seemed to like his new life being captain of the ship and being part of the search. Since their discovery of the small deposit of gold bullion, he was determined that he personally would be the one to find the rest of the *Atocha*.

The name *Saba Rock* was vaguely familiar to me, but I couldn't place it at first. Though Ed and Susan hadn't been aboard all that long, they were able to give us enough information about the ship's recent past that the light bulb came on. This was the same boat offered to us over the telephone for that bogus expedition last summer. Even more of a surprise was the owner. Don Durant, the diver who had left Salvors not long after the **Margarita Main Pile** was found, had convinced his father that treasure hunting was a far better use of ships than was normal commercial work. Durant Senior owned the *Saba Rock*.

When its intended first expedition didn't pan out, a new project and crew was needed. Perhaps the oversize blowers would be just the ticket in the *Atocha's* deep sands. About the same time as agreements were being made for the *Saba Rock* to work the *Atocha*, Ed happened to be in the right place at the right time. After a chance meeting in a bar, he became captain of the ship, ready to take his turn in the long line of subcontractor efforts.

It all seemed a bit surreal. In my experience, retired jocks and their cheerleader girlfriends had never been drawn to our unusual vocation. Susan and Ed were the physical manifestations of a P.R. person's dream team, looking as "All American" as one could get. Here was this handsome guy and his supportive girlfriend running the largest, most powerful vessel on the site. With no prior ship handling or salvage experience, they had even been lucky enough to find some notable artifacts.

After the news of their find had spread, reporters regularly visited the ship to document Ed and Susan's activities. KT and I watched a photographer snap pictures of bikini-clad Susan on the back deck while she trimmed her toenails. She intimated her ignorance about the gold bars she had found while diving in one of the *Saba Rock's* excavations. She said she thought they were plastic.

Ed and Susan were socially friendly but Ed wasn't particularly interested in my advice. After finding the gold bullion, he decided that his own instincts were somehow more acute than that of any Treasure Salvors employee. Perhaps the media attention had tickled his ego. He sometimes proclaimed that finding the *Atocha* would make him the most famous treasure hunter of all. I could never tell if he was kidding or serious.

Ed had taken to digging blocks of holes in a shotgun attempt to relocate more of the *Atocha* trail, and they had all turned up empty. This far southeast from the **Nine Cannons,** the artifact track was just about invisible. The *Saba Rock's* power was impressive; it could punch a massive hole through eight to twelve feet of hard mud straight to the bedrock. Our company boats were lucky to dig down a foot and a half in the same place.

Pushing further to the east, the *Dauntless* and *Swordfish* leapfrogged each other, diving on anomalies originally logged by the *Plus Ultra*. Almost always the hits were found to be modern junk or bombs, but against all odds an *Atocha* era spike or barrel hoop fragment would sometimes turn up. As they moved further along, their "trail" had become so ill defined that it was virtually no trail at all. Frustration and excitement ran at a dead heat.

Life for me aboard the *Saba Rock* quickly grew stale. Since Ed brushed off my suggestions there was little for me to do except dive empty excavations with KT. In Australia she had suffered a near fatal diving accident due to an equipment malfunction in the worst possible conditions. KT bravely decided to continue her diving career but her confidence had been bruised. It was nice to dive with her while she worked through her demons.

Summer had not quite arrived, but the temperature was already un-

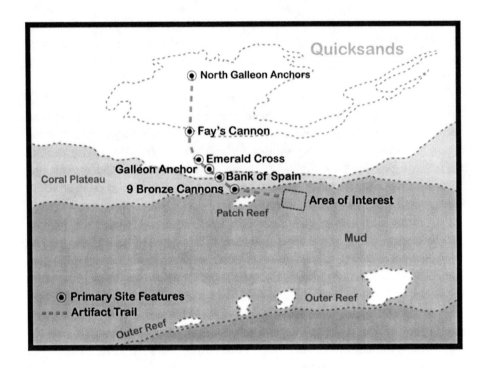

comfortable. The *Saba Rock's* rusty, exposed back deck acreage was Hell's front porch in the direct sun; the ship's gunwales effectively reflected the heat inward over the already broiling surface. Adding to the treat, one of the crew had accidentally pumped all of the ship's potable water over the side, so now the only source of fresh water was collected air conditioner drippings. The perpetually running air conditioners produced just enough water to wash dishes and to quick rinse diving regulators but little else. We were grungy, covered with salt and sweat. With the exception of cool living quarters, life had become a recant of the *Arbutus*.

Captain Ed began gradually moving the ship's digging patterns more towards a presumed intersect of the shadowy scatter trail. The *Saba Rock* continued to dig each excavation right down to the bedrock. The very few artifacts found by the *Dauntless* and *Swordfish* had been located only a couple of feet down in the mud, but none of us knew for sure how deep the artifact strata really was in the sediment column. Were the company boats not able to probe deeply enough to find heavier artifacts, or did the weighty gold bars found in the bottom of the *Saba Rock* holes fall there as the surrounding overburden was blown away? It was a big mystery.

Memorial Day 1985 was another blazing hot Gulf Stream day. KT and I had explored the last two empty excavations together. To avoid decompression sickness from diving in this deeper water, there were rigid restrictions on how much time we could spend on the bottom. Both of us had just enough time to quickly check out the next hole before we would have to stay on deck for another two hours.

When the next excavation was completed, we entered the water with our metal detectors and headed downward into the dark swirl. The current here was much less intense, and it took some time for the displaced sediment to be carried away by the tidal flow. We both worked in the grainy twilight on opposite sides of the excavation. I found nothing on the exposed bedrock or imbedded in the mud wall on my side of the hole and gradually worked my detector search upward.

The shape of the excavation was like an inverted bell, near the seafloor's original surface the hole broadened and dished out. Around the very rim of the excavation little more than a foot or so of the mud had actually been removed. I ran my detector back and forth across this zone as I swam around the perimeter. The previously silent detector suddenly gave a quick, strong signal.

I used my knife to section out and remove a chunk of the mud and then fanned my hand over the spot. A pottery shard became visible; its lead glazing had set off the detector. I gingerly removed the fragment and passed the detector again over the spot to confirm I had recovered all of it. The detector rang out again. As I fanned more over the growing hole, KT swam up to see what I was doing.

I knew I was running out of air. Minutes earlier an inspection of my tank's pressure gauge had showed less than five hundred pounds. The regulator was starting to draw hard with each breath. I pushed my hand into the mud, grabbed something hard and gently held the object up in the clearer water. Some of the sediment clinging to the item smoked off in the current. I could clearly see five cut emeralds held in some sort of gold jewelry. I showed the piece to KT and pointed to the spot it had come from before heading to the surface with my prize.

My hand grabbed the dive ladder just as the SCUBA tank ran out of air. I guessed that this was another isolated deposit of artifacts, everything found on this mud trail so far had been very widely scattered. My animations on the back deck attracted Captain Ed and some of the others. They took turns fingering the mud-clotted jewelry. As I caught my breath enough to describe the nature of the find, KT climbed up the dive ladder and presented four more pieces of gold and emerald jewelry similar to mine, found in the same spot. She too had run out of air and bottom time, but was convinced that there was still more down there.

The *Dauntless* was working on a hit not far off our stern and her crew could plainly see the commotion on the *Saba Rock's* back deck. Ed confirmed

Still wet and packed with mud, this emerald jewelry was photographed just after being brought to the surface.

our find to the *Dauntless* via radio and soon the Key West office knew as well. Mel was elated to hear about the discovery, and decided to come out to the site for a personal inspection as quickly as he could.

It was time for us to get back to work. Treasure Salvors' company procedure in a situation like this was to immediately mark or buoy the spot and plot the coordinates. The whole area would need to be minutely examined in a controlled, fine scale inspection. The *Saba Rock's* blowers were too powerful; we would need to rig up an airlift for a gentle, non-intrusive excavation. This procedure was so ingrained in me that I was caught off guard by Captain Ed's actions.

Ed fired up the winch engine, obviously intending to winch the ship away from the spot. I panicked, and over the noise of the diesel challenged his intentions. We hadn't yet marked the location in any way. The Del Norte system T.S. had lent the *Saba Rock* couldn't be used to record the position since the unit's antennae had been mounted too far forward of the stern. It had also been damaged and was not working reliably. I insisted that Ed have a diver immediately put a buoy on the spot, it would be our only reliable reference if he moved the boat.

Captain Ed was really cranked up on adrenaline, ready to act and in no mood to listen. He told me that he wasn't interested in picking up a few trinkets - he was here to find the whole *Atocha*. It absolutely **had** to be nearby. With that, he engaged the winch and began dramatically shifting the ship's position, apparently directed by an internal Ouija.

KT and I quickly prepared a buoy and tossed it over the side in a last second attempt to mark the jewelry's general area. Our effort was foiled when one of the *Saba Rock's* now slackened anchor cables fouled the buoy and dragged it off.

All of our subsequent excavations came up empty. Mel and entourage arrived the next day via *Bookmaker*. Never one for formalities, Mel rebuffed maritime protocol by ignoring Captain Ed to talk to me. I reported the nature of the emerald jewelry find and what actions Ed had since taken.

Mel wasn't impressed that the *Saba Rock* had moved away from the find's location. He wanted the ship immediately put back on the emerald jewelry's position so an airlift search of the area could take place. The problem was that Ed had jinked the ship's moorage around so much that he couldn't find the spot again. With no buoy or Del Norte coordinates to go by, the *Saba Rock* had once again lost the trail.

The *Dauntless* ended up providing the day's excitement for Mel in lieu of the irretrievable jewelry location. Working a few hundred meters off the *Saba Rock's* stern, divers found an aged timber buried in the mud. Its time worn shape had a deep groove suggesting damage by repeated rope movement

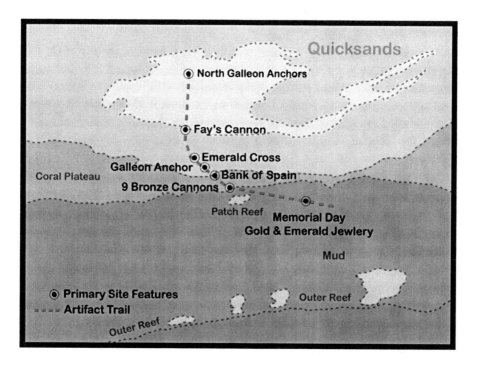

across its surface. Without analysis it would be impossible to prove it was from the *Atocha,* but its discovery was enough to excite Mel and the *Dauntless* crew. Mel returned to Key West later that day, eager to promote our latest finds to potential investors and the press.

The emerald jewelry area would now become known as **Memorial Day**. Though other items had been found between the **Nine Cannons** and this latest feature, the emerald jewelry deposit seemed final confirmation that the *Atocha's* main pile must exist further to the east instead of the southeast. The long held theory that the galleon had hit and sunk near the **Patch Reef** was finally disproved.

After the **Memorial Day** find, the *Saba Rock's* luck ran out. Completely frustrated, Captain Ed aggressively winched the ship back and forth, digging here and there. He didn't suffer alone, after weeks of this everyone aboard was cranky. I was really in a black mood; all the random, empty holes and my inability to affect any change in Ed's methods were psychologically exhausting. Doubly worse was that there was no definitive plan for us to rotate off the damn ship.

A short break from the tedium was supplied by a visiting reporter with his own interesting stories to tell. Jedwin Smith wrote for the Atlanta Constitution and had lived a daring life as a foreign war correspondent. He became enchanted by the *Atocha* story and was particularly interested in the personal stories of the participants. Jedwin saw the expedition as a larger story than just one article, he would return again and again over the years to document the progress.

As the *Saba Rock* groped blindly trying to relocate the trail, the *Dauntless* and *Swordfish* were making slow progress exploring ever eastward. They ranged further and further away from the *Saba Rock*, every now and then finding another small spike or other tiny artifact that encouraged yet further eastward movement. Not bothering to cover every square inch of ocean floor, they leapfrogged each other's position investigating any easterly *Plus Ultra* mag hits on file.

After nearly six weeks aboard the *Saba Rock*, KT and I were at last relieved, returning to port to enjoy a few days off in Key West. Ted and Leah had graciously given us their downstairs "mother-in-law" apartment, further strengthening our ties to 417 Cactus Dr. While we relaxed and caught up on shore bound responsibilities, Tom Ford made a fortuitous mistake.

Recently on weekends, Mel and office cartographer Ed Little would sometimes review the charts of plotted *Plus Ultra* hits. They looked for anything remotely interesting, especially if it projected more to the east. The Del Norte coordinates of any anomaly of interest would be radioed to the boats out at the site.

The iron swivel gun found laying near other artifacts on the sea floor
Photo: Tom Ford

Aboard the *Swordfish*, Tom Ford routinely wrote down the position of the latest target radioed to him from the office. By accident, he confused the coordinates from the two Del Norte towers. He moved the *Swordfish* to the spot indicated by the faulty numbers and had his crew throw a buoy on the spot. It was a huge jump beyond where they had previously been; the confused coordinates had moved them eastward more than a mile.

The *Swordfish* divers got a surprise not far from their buoy block. A 17th century swivel gun was lying on the bottom. It was a common anti-personnel firearm that was usually mounted on or near the railing of a ship. The *Atocha* was known to have carried several of them. Scattered around the small cannon were groups of chain shot; two cannon balls chained together. This ordinance was designed to take out rigging and sails on an enemy ship. There were even the remains of a small barrel or cask, its hoops and some of the wood were preserved in the sediment.

Tom's unintended slipup quickly advanced the eastward search to the area around this **Swivel Gun**. The *Dauntless* quickly moved into the area and all the company boats began exploring more unsearched anomalies logged in this area as fast as the office could radio them. Eventually several hundred yards to the northeast of the **Swivel Gun** more barrel hoops and ballast stone was found.

The mud corridor search, which had always seemed so tenuous, was quickly coming into focus for the boat crews. Even though archaeologist Duncan Mathewson boldly declared that the ballast was likely from the *Margarita*, it was a theory solidly rejected by everyone else in the operations. WE all knew we were getting close, we could **feel** it.

At the end of the **Swivel Gun** discovery trip, Tom Ford stopped by Ted's house to report on this latest development. KT and I listened in as he described the artifacts found in the **Swivel Gun** area. He must have known how desperate we were to be back in the game, sitting here in port as unassigned crew was torture. Before the evening was out he asked KT and me to join his crew aboard the *Swordfish*.

It was the kindest possible gesture, and held potential hassles for him. I had captained the *Swordfish* longer than Tom, had enjoyed great success and was more fluent in the boat's operation. Years of being captain had strongly affected my personality; habitually I took charge of situations. Obviously that would be a big problem in this situation, there could only be one captain, and he was it. I would have to try hard to become more passive. We asked Tom why he had extended the offer to which he graciously replied, "Because you both deserve to be there."

From that point on we were back in the mix. Not long after we had returned to the *Swordfish*, another strange turn of events played out. The *Saba Rock's* contract to work the *Atocha* site ran out and Mel was not currently in the mood to renew it. The ship was moved off site and became a rust streaked Steel Island, anchored outside of territorial waters with uncertain prospects.

Captain Ed and Susan suddenly found themselves unemployed and were ferried back to Key West. Like the rest of us they could sense that we were closing in on the *Atocha*, and unexpectedly now found themselves on the outside looking in. They hung around our docks for a while, trying to get onto our company boats as crew. They found little sympathy and dejectedly faded into their former lives.

The *Swordfish* and *Dauntless* had been joined in the search by the recently recommissioned *Virgalona*. R.D. LeClaire had convinced Mel that he should reactivate the old boat and become its captain. I was pleased to see that Curtis White was serving as first mate. Curtis had started with Treasure Salvors as a carpenter not long after I had first been hired. Since we had both come on the scene about the same time there was a natural connection between us. After years of working on "the beach," he had finally got his dive certification and was now getting the chance to participate in the offshore search first hand.

KT and I had some friends living on the west coast of Florida who had once told us that if we ever felt we were going to find the *Atocha* that we should contact them. We called them as well as my parents, telling them that the time to invest was now as the discovery seemed imminent. None chose to take our advice.

July 19, 1985 was the last day of a long trip aboard the *Swordfish*. For days we had checked out magnetometer anomalies along with *Virgalona* and *Dauntless*. After finding nothing more around the **Swivel Gun** feature, a sprinkle of artifacts several hundred meters to the east-northeast was located. *Swordfish's* plan was to go into Key West that evening to refuel and reprovision, returning the next day with supplies for all the boats.

All three boats had been advancing eastward while exploring more "hits" in the same general area. Back in Key West, Mel and Ed Little had noticed a correlation between a large mag hit and a low rise on the sea floor recorded on a *Plus Ultra* sidescan sonar printout. The anomalies themselves were unremarkable, looking just like the thousands that we had searched before.

The *Dauntless* had set up to dig near the coordinates, and began to find artifacts in increasing numbers. As *Swordfish* steamed for Key West in the early evening, *Dauntless* reported that they had found some coins and even copper ingots.

Early the next day we were back on the site. After distributing supplies to the other boats, we got back to our own exploration. The *Dauntless* had stayed anchored for the night over the spot they had worked the previous day, and was continuing their excavations. *Virgalona* had moved to explore a small patch reef about three hundred meters east of the *Dauntless'* position. Tom Ford decided to set up the *Swordfish* just to the north of the *Dauntless*, and began our excavations there.

We dug a number of holes that morning but found nothing. The bottom here was a gooey silt, overlaying harder mud beneath. There were a few stunted looking sponges and little else to see. The water depth here was nearly sixty feet, limiting the time each diver could work under water.

The *Virgalona* hadn't found anything either. Only the *Dauntless*, straddled over the narrow artifact trail, reported continued finds. My own instincts told me that we needed to move the *Swordfish* to the south of the *Dauntless* and look there. As the day wore on I bit my tongue often while Tom patiently dug one empty hole after another. Our crew's dive time was being eaten up with little result. I was the last diver aboard left with any real bottom time remaining.

Satisfied at last that we were in the wrong spot, Tom decided to shift the *Swordfish's* position. Like me, he had come to the decision that our next spot to check should be to the south of the *Dauntless*. We had all come from the west, the *Virgalona* was finding nothing to the east and *Swordfish's* excavations were empty to the north. After the passing of a brief squall, we quickly picked up our anchors and moved about one hundred feet south of the *Dauntless* and set up again. It was a good placement, not too far, not too close. I prepared my dive gear, intending to do a brief "recon" dive on the seafloor beneath us before we began excavation.

Under the *Dauntless*, Greg Wareham and Andy Matrocci worked their latest excavation. Like Don Durant years earlier on the *Margarita*, Greg decided to explore on his own outside the hole. He had once made a living recovering golf balls out of golf course water hazards. Scooting the detector ahead of him some distance from the *Dauntless*, he made the discovery of a lifetime. Greg quickly swam back toward the *Dauntless* to get Andy.

I stood at the top of the *Swordfish's* dive ladder, about to jump in when two divers broke the surface about ten feet away. It was Greg and Andy, who shouted above the diesel engines "It's all here, it's right here!"

I took the regulator out of my mouth. "What's there?" Everyone aboard the *Swordfish* was instantly at the rail.

They yelled back. "The **Main Pile**, it's right here, beneath you!"

I was already in the water, detector in hand, streaming bubbles, going straight down. The water column grew more clotted with turbidity the deeper I went; the tide carried sediment from the *Dauntless'* excavations through the area. The jade green color of the water had turned nearly grey by the time I hit the bottom.

I knelt on the seafloor for a second to let my eyes adjust to the dim light. At first the mud plain before me appeared unremarkable, but after swimming a few yards I could see a change. A patchy colony of small soft corals and sponges emerged, hinting at a change in the bottom configuration. I looked very closely. Despite centuries of sedimentation and exposure, there were faint, brick-like shapes the size of silver bars lying askew like dropped pick-up sticks.

From nooks between the bars, small lobsters waved their antennas at me. There were far more fish in the area than was usual in the normally vacant mud area. Fishermen had been to this place previously; there were old broken strands of monofilament fishing line snagged among the bottom features. They had missed the catch of the century. I waved the detector over the bar shapes and it screamed. So this was the fabled **Atocha Main Pile** at long last!

I swam further to see what else was visible. Numerous shapes of other divers flickered quickly by like shy ghosts. Lack of bottom time or not, everyone wanted to see it for themselves. Above me I could hear propellers and engine noise working the water.

The visible artifact mound wasn't very pronounced; it protruded less than a foot above the surrounding mud. The mound seemed to be oriented almost north and south and was perhaps twenty feet wide by sixty feet long. At the southernmost end were the silver bars, and to the north greenish copper ingots lay exposed amongst the tumble of algae and sediment softened ballast stones. An unexploded five hundred pound World War II bomb lay here, the probable source of the mag hit. There was no telling at this point how settled into the mud the artifact mass was, but there was little doubt this was just the tip of the iceberg.

I surfaced into confusion and activity. Kane had already winched the *Dauntless* just a few feet away from the *Swordfish* and the *Virgalona* was now anchored up close to our other side. Each vessel had amplified loudspeakers on its back deck that broadcast all radio traffic between the boats and the office. If one boat transmitted, the other's radio equipment received both sides of the conversation so everyone could hear it. Kane proudly called in to the Office "Put away the charts, we've found it!"

An immediate and involuntary reaction spread through the crews. After the many long years, the fatalities, sacrifices, thousands of empty holes, and the near terminal financial and legal bleed outs, it was imperative that each boat have on its deck some physical confirmation of the event. This was our moon landing, our Mount Everest and we had to plant the flag RIGHT NOW. The *Atocha's* vast cargo of silver bars had always represented our vision of what the main pile might look like. In an emotional spasm, KT and J.J.

Shapes of silver bars are visible protruding out of the mud.

Bettencourt brought up one of the eighty-pound bars for the *Swordfish*. Divers from the *Dauntless* and *Virgalona* also staked their claim.

With so much high-energy animation going on around the boats, I only processed brief mental snapshots. *Swordfish* diver Paul Busch, who had served with KT and me on the *Saba Rock,* used his underwater movie camera to film the undisturbed **Atocha Main Pile**. *Dauntless* crewman Mike Meyer announced he had spotted gold bars lying exposed on the mound. Two *Dauntless* divers stood on deck skipping previously hard won "pieces-of-eight" across the water's surface. It was an extravagance inspired by what lay beneath them.

It suddenly occurred to me that with all the new faces, Kane and I were the only T.S. crew that had been on hand for the discovery of both the **Margarita Main Pile** and the **Atocha Main Pile**. The *Swordfish* and the *Virgalona* could claim the same honor. A more surprising fact was that it was ten years to the day since the *Northwind* sinking and the deaths of Dirk Fisher, Angel Fisher and Rick Gage.

There was nearly an hour of freeform deck and underwater activity, boisterous radio calls and backslapping. We finally sobered into a more unified purpose. Both the *Swordfish* and the *Dauntless* would use their blowers to very gently dust off the top of the mound to get a clearer picture as to what was there. Test pits could then be dug around the perimeter of the "pile" with airlifts to determine the actual dimensions.

In this deeper water, our boat's mail boxes lost much of their ability to actually dig holes but were perfect for brushing away the most loosely held sediments. The *Virgalona's* underpowered blowers had no real effect here, and with no airlift compressor of her own, was relegated into being a diver support vessel.

As Kane repositioned the *Dauntless* over the silver bars, he inadvertently "shoved" the *Swordfish* and *Virgalona* a little further south. We didn't realize this until checking out the results from our first mutual "dusting" operation. *Dauntless'* blowers had fanned the bar area and *Swordfish* had uncovered a large intact olive jar nestled amongst a wooden structure that had not been visible before. We had never imagined that any of the *Atocha* itself would have survived - how much more of the actual ship might be here?

Tom, KT and I held a pow wow. The Del Norte navigation system wasn't accurate enough for the detailed mapping this feature would require. We needed to be able to measure artifact relationships horizontally and vertically in centimeters rather than meters. The PVC grids Tom had on board were small, perfect for documenting individual artifacts such as we usually found but not a large, complicated, high-density area like this. As the "archaeology crew", we needed to quickly come up with a plan. The pressure for recovery operations to begin immediately was intense, especially on the *Dauntless*. Kane's destiny was in recovering treasure, not recording shipwreck data. His crew was eager to help him.

I suggested a simple, mechanical version of the Del Norte measuring system. We could survey in a series of metal stakes, equidistant and running north-south along the axis of the mound. Each stake would have the end of a tape measure attached to it and the tapes from two or more stakes could be used to give the distance of each artifact to the surveyed stakes. We could also use a bubble line fixed to our stake system to give vertical dimensions. Tom liked the idea, so we set about to implement it.

Near the southernmost boundary of the exposed wood we hammered our first steel stake benchmark named "A" into the caprock. From that point we went due north, locating three more named "B, C, and D," spaced thirty feet from each other and connected by a heavy nylon twine. The ends of survey measuring tapes were attached to each stake at the same elevation. We were ready to go.

It was not a moment too soon. The *Dauntless* divers began to recover exposed silver bars, loading them into plastic milk crates attached to ropes leading to the ship's winches. The *Virgalona* guys were eager to get into the action too. As the recovery divers attached artifact tags to the bars, we measured the positions and wrote the data with grease pencils on writing slates. The rate of recovery began to accelerate, and Tom's protests to Kane barely moderated

Setting up the mapping benchmarks on the *Atocha* **"Main Pile"**
Photo: KT Budde-Jones

Loose silver pieces-of-eight scattered around a silver bar

the pace. On deck, I tried to explain why it was important to document what we were finding to a smiling newbie *Dauntless* crewmember who quipped, "Fuck archaeology… it just slows things down!"

Tom was in a tough spot. He had unofficially assumed the mantle of seaborne operations manager when Bob Moran departed, but his authority didn't necessarily apply to any of the Fishers. He was a student of scientific methodology and felt obligated to see that the recovery proceeded in a deliberate, highly documented fashion. Tom could direct the *Virgalona* crew in a formal dissection of the *Atocha's* remains but how could he mediate the same with Kane?

Kane didn't worry himself over academic concerns. He had been trained and praised his whole life by Mel to find and recover treasure. His family's stubborn obsession to find the *Atocha* had cost them dearly in so many ways. Bringing it all up now was his righteous payback.

Tom played the weak, but only hand he had. He called Mel on the radio, suggesting that Mel might want to warn his crews about recovering material until a proper archeological plan was agreed upon and archaeologist Duncan Mathewson had visited the site. Mel in his proud, fatherly way gently cautioned Kane about "Bringin' up too much right away".

We spent the rest of the day measuring in more silver bars that the *Dauntless* divers continued to bring up. As the weighty ingots came to the surface they had to be stored below decks to keep from making the boat top heavy. Both *Swordfish* and *Virgalona* had some recovered bars on their decks too. Using a waterproof opaque plastic charting film, I began to plot all the artifacts that so far had been brought up.

That night the three boats stayed anchored in a broadened three abreast position. There was no more shouting or carrying on, in fact the mood six stories above the *Atocha's* ballast pile was subdued and introspective. I could not say what the other crews thought, but the primary thought aboard the *Swordfish* was now that the shipwreck had been found we would all soon be out of a job. After the years of searching, the actual discovery had seemed almost painfully abrupt and final. I was surprised to find myself personally unprepared to achieve our goal.

I did not imagine how in this new phase the *Atocha* project would become like an underwater construction site. Probably no one in the company understood that our expedition had just turned into a business.

# TRIBBITS

The next day Mel, Deo, Ted, Leah, the office staff, friends, numerous reporters and camera crews descended on us in a small boat armada that appeared to have emptied all of Key West's docks. Even Donnie Jonas brought out the *J.B. Magruder* and rafted it up to our company boats.

The influx of excited, normally shore bound members of Treasure Salvors' staff brought with them new gossip that spread quickly throughout the boat crews. Mel had been out of the office when the message about our discovery had first come in. Local radio stations had broadcast an announcement for Mel that the *Atocha* had been found and that he should return to the museum immediately. There was also a rumor that Mel might be buying the *J.B. Magruder* to help in the recovery. The media was storming the office, trying to get reporters out to the site to cover the white-hot story.

Mel and Deo came to see the remains of the *Atocha* and its cargo. Since Mel no longer had dive gear of his own, I gladly lent my tank, regulator and buoyancy compensator to him while other crew members helped Deo gear up. Mel and Deo swam off hand in hand over the **Main Pile**, a spiritual and physical culmination of their own sixteen year long tragi-comedy. I wondered what they must have been thinking; their lives had been committed so completely to this one goal. Against all odds the project had succeeded, but success exacted a price. Mel and Deo themselves held some of the largest receipts.

It was hard to control all the diving activity that day, so many people with little diving experience wanted to see the wreck. Our dive boat crews turned into lifeguards as more than a few rookies were carried off by the tide

Jimmy Buffet sings while sitting with Mel Fisher on a bench made of silver bars aboard *Magruder*.

or got lost navigating the faded world below. By coincidence, singer Jimmy Buffet happened to be a few miles away, using the awash wreck of the *Arbutus* for an album cover photo shoot. He heard of the discovery of the *Atocha* and eventually made his way over to the huge raft of boats hovering over the galleon's grave.

Due to stability concerns on the company boats, we had been transferring the heavy recovered silver bars to the *J.B. Magruder* from the *Dauntless*, *Swordfish* and *Virgalona*. The *Magruder* would be heading back to Key West at the end of the day. We stacked up the bars in a small pile on the *Magruder's* back deck. Jimmy joined Mel in sitting on this bench of silver and entertained us with an impromptu solo concert of "Margaritaville". Not much real work happened on the site that day; it was a time for everyone to unload years of frustration.

A photographer from Time Magazine gathered the boat crews and Mel together for a game-day portrait on the back deck of the *Dauntless*. We all hoisted champagne glasses as the shutter clicked. For that one moment we were all as one; the petty bickering, jealousies and competition between us temporarily forgotten. Many of us had invested years of our lives in this effort and for the time being we blended together in the celebratory afterglow.

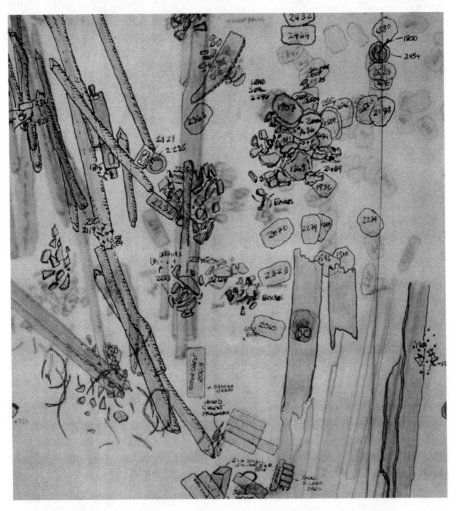

A very small portion of the overlay **Main Pile** site map. The fainter images are on underlying layers. Some of the items visible in this view include dunnage, copper ingots, ceramics, coins chests, silver bars, orlop deck planking, silver bowl, candlestick, barrel hoop fragments, and a cannon carriage axle.

At the end of the day the fleet of boats heading for Key West finally departed, leaving the three company boats riding easily over the *Atocha's* ballast mound. We were all exhausted.

The next few days were a conveyor belt of charter boats bringing more reporters and hangers-on. Despite the distractions, the artifact mapping procedure gradually tightened up. Though the *Swordfish* divers were still doing the bulk of the underwater measuring, the other crews began supplying some of their own data as well. I had set up my mapping station in the *Swordfish's* wheelhouse and spent all of my surface time carefully plotting in each artifact as the information was turned in. Without really thinking about it I had assumed the complete responsibility of mapping all of the artifacts found on the site. If this chart was to be the only detailed site plan produced on the **Main Pile** I was determined that it be dead accurate.

The focus of my own underwater mapping efforts was the wood structure. As the airlift teams gently worked into the ballast pile more and more of the wood became exposed. The timber represented a complicated mapping problem; each piece required numerous measurements to record contours, notable features and elevations within the artifact strata. Each piece was assigned its own fluorescent red identifying letter. Underwater I personally rechecked each of the hundreds of measurements at least five times before committing the information to the site map. By using several overlays, on paper I could represent artifacts found at different strata depths.

The volume of ballast stone was beyond our experience and presented quite a problem. It had to be removed, one stone at a time by hand. Nearly two hundred tons of rock were intermingled with the other *Atocha* artifacts and it had to be put somewhere. Teams of divers dug test pits with airlifts in the shallow mud around the periphery of the ballast mound, looking for areas without artifacts. If a spot was found to be clear of *Atocha* material, the hole would be used as a ballast "dump" to place the stones as they were being removed. Eventually twelve ballast dump locations surrounded the **Main Pile**.

I put aside a medium sized ballast stone for myself and for a few days soaked it in fresh water. When it dried, I had all the crew members from each boat autograph it, eventually including Dr. Gene Lyon, Mel, Deo, Ted, Leah and Duncan Mathewson as well. It was to be my personal keepsake of the finding of the *Atocha*, a literal touchstone.

Archaeologist Duncan Mathewson and conservator Jim Sinclair at last arrived on the site with yet another convoy of media chartered vessels. They brought with them materials to set up a large grid system commonly used on high artifact density recoveries. So far the size of the actual area of our excavation on the **Main Pile** was actually quite small. Exposed artifacts in the undisturbed northerly silver bar area looked like a good spot to begin estab-

One of the ballast dumps created by hand moving the heavy stones aside while excavating the site. Photo: KT Budde-Jones

lishing the grid system. We could easily tie in the position of this new grid to our already surveyed benchmarks.

Recoveries were curtailed until the grid work was in place, then the airlifts were brought in to begin excavation in the new spot. Almost immediately we began finding the remains of crushed wooden boxes that had once held silver coins. The coins themselves were frozen in the positions of their spill, fixed in place by concretions of silver sulfide. The crushed coin chests were photographed, tagged, measured in and removed. Completely intact wooden coin chests were found buried further down in the mud as the excavations deepened. Here as in dreams, real treasure chests!

The chests were actually simple rectangular boxes made of red cedar, nailed together with no locks or hinges. The small iron nails were found to be completely eaten away by corrosion and the wooden boards of each chest could be removed without damage by gently working a putty knife along the seams. The chests were heavy; each weighed about one hundred and seventy five pounds. It was important to remove the delicate waterlogged wood before trying to move the weighty lump of coins. The wooden components from each chest were placed together in plastic bags and brought up with the coins they held. Conveniently the coins were fused together into a homogenous mass by a crust of silver sulfide concretion and could be recovered as a single item.

The *Dauntless* divers made an interesting discovery amongst the coin chests. Silver artifacts of all sorts were found packed in a very large silver bowl. There were spoons, forks, plates, incense burners, candlesticks and cups. Along with the other silver pieces were two small silver pitchers, one with an "A" standing proud on the lid and the other with a "V". They were a portable aqua/vino set used for communions. The pitchers and other silver items still had traces of the brown paper wrapped on them when they were packed for their trip. It was amazing how effective a preservative the mud had been.

The grid system quickly proved unpopular as the divers moved heavy artifacts through or around it. Though well anchored, it was always being knocked out of place as the divers tried to work against the tides with the airlifts or while moving the heavy ballast stone and coin chests. Lots of effort was expended in continually rechecking its necessarily exact positioning, only to have it bumped askew again. Eventually everyone agreed that our original

From top: KT, Bill Barron, Paul Busch and Jim Sinclair examine several silver bars that have just been winched to the surface.

measuring system was just as accurate and far more practical since the measuring tapes and bubble lines could easily be moved aside when it was time to extract an artifact. The grid system was removed and abandoned.

The airlifts we used were six-inch diameter PVC tubes forty feet long with a valve mechanism for diver control at the weighted bottom end. Air lines snaked down from the shipboard compressors to the valve, allowing the operator to adjust the amount of air being pumped into his tube, increasing or decreasing its suction power. Even against the tide's force the airlifts stood straight up and down in the water column, held firmly in that position by the rising and expanding air within them.

Instead of using the mouth of the airlifts to dig directly into the mud, we held the suction end a foot or two above the bottom and used them to suck up the disturbed overburden that we hand fanned away. In theory this was to keep lightweight or small artifacts from being sucked up the tube. The silt the airlifts picked up was expelled forty feet above us and was carried with the tide like smoke from a smokestack. The mud itself also contained millions of small shells and shell fragments, which being heavier, gently floated back down on top of us. With four or more airlifts going at the same time the sensation on the seafloor was similar to being in a twilight, soft flaked snow storm. It was a very calming sensation, especially when

A freshly excavated intact coin chest with its lid removed   Photo: Tom Ford

A diver gently excavates amongst the Atocha's lower hull structure.

compared to the heat and noise six stories above.

Once in a while one of the divers carelessly got the airlift suction end too close to a ballast stone. If a large stone was sucked against the mouth, the airlift quickly turned into an ICBM missile, blasting straight to the surface from the tube's sudden uncontrolled buoyancy. This was dangerous for two reasons, if the operating diver was carried rapidly upward with the airlift he might embolize or get the bends. Boats or swimmers on the surface could also suffer severe consequences from the impact of a runaway airlift's rocket-like ascent.

Less dangerous but still worrisome was the dull rattling sound of a small rock being carried up someone's airlift. That sound meant that in a few seconds that stone, perhaps weighing five or ten pounds would soon be falling back down on one of us when it cleared the upper end of the airlift tube.

Like many of the other crewmembers, I was mystified by archaeologist Duncan Mathewson's somewhat tardy arrival on the site. Had he been on the site in the first days, his influence would have been useful in dampening the initially over-aggressive rate of artifact recovery. As it was, the *Dauntless* crew still continued to far outstrip the other boats in items brought up, leaving us with most of the measuring and mapping details. Had Mel set a bonus for the most artifacts brought up? Kane and his crew surely acted like it.

The *Swordfish* divers assembled around the port wheelhouse door for a serious heart to heart with Duncan. We had worked very hard to document and precisely map the **Main Pile** feature. Our insistence on an archaeologically prudent recovery had spawned some unwelcome friction between us and the *Dauntless* crew. Mel had commented appreciatively about our efforts to record the recovery, but he seemed much more enthusiastic about the ever increasing pile of silver bars and coin chests being brought up by Kane. With no forthcoming directives from Duncan, the company was simply living up to its name - Treasure Salvors.

Duncan told us he had been busy trying to organize a team of contract archaeologists who would soon be involved in daily activities on the site. There were valuable studies to be made of the **Main Pile** area and he needed a lot of help to do it. He stressed that the site mapping and data collection work we had done so far was crucial. His immediate plan was to hold a meeting amongst all the boat crews to assure that everyone was on the same page.

All work was shut down while Duncan spoke before the three assembled crews on the *Dauntless'* back deck. I couldn't tell if he was speaking more for the benefit of the cameras recording his speech, or our impatient employees - many of whom saw this pep talk as little more than a media photo-op. The impact of his words about the importance of proper archaeological procedure was difficult to measure when looking at the collection of fidgeting or impassive divers.

Duncan's opinions hadn't always been taken so casually. He had been the first person to scientifically examine Treasure Salvors' early discoveries near the **Bank of Spain** and hypothesize that the majority of the wreck was in deeper water. Dirk Fisher had been intrigued by the concept, which had led him to move *Northwind* toward the seafloor drop-off into the mud. It was there that the nine cannons were found just days before the *Northwind* sank. Finally proving Duncan's theory would take another ten years and over five nautical miles of movement further into the deeper water area. Whatever dissatisfaction we felt toward him over his absence didn't alter his contribution to our triumph.

After Duncan and the attendant media left in their boats for Key West, things pretty much went back to the way they had been. There were still coin chests left to measure, photograph and excavate. Airlifts went back into action, staining the surface water with mud carried up by the fountains of rising bubbles.

One of Kane's "posse" served aboard *Dauntless* as a deck hand. He couldn't pronounce the word turbidity to describe the milky water, so he substituted his own word - "tribbits" instead. It was a good word to describe not only the opaque water, but also our own lack of clarity as a recovery team.

Mother Nature provided a chance for all of us to exhale. Hurricane Bob approached the lower Keys near enough to chase us all off the wreck site and back to the safety of Key West. The break in sea operations gave us a time to rest, regroup and collect some of the new equipment that Ted had procured, including specially designed lift baskets for artifact recovery and waterproof paper and pens for underwater data collection. While in port we were able to witness first hand the incredible volumes of people that now stampeded the museum. For the next several years visitors would watch in amazement as we regularly unloaded pickup truck loads of artifacts from the boats.

One notable change on this main pile was my status with Don Kincaid. He was no longer involved in any of the seaborne operations but was still on hand for photographic and P.R. purposes. In a way he still represented the original crew who had first sought the *Atocha* so many years ago, but by now I too was an old hand with lots of hard earned experience under my weight belt. Any animosity between us was long gone, over time we had come to mutually respect each other's accomplishments. KT and I admired both his sensitivity to the ocean's degradation and his statesmanship when talking about the *Atocha*.

We returned to the **Main Pile** after the storm's passing with an improved routine. All three boats began sharing data acquisition a little more equally, and the original, near desperate recovery rate ratcheted back. The continuing daily rotation of media representatives ferried back and forth between the site and Key West had become unremarkable, except for one towering exception. A National Geographic film team came to stay with us for several weeks. In all the diver's minds, this was as good as it got.

To us, National Geographic coverage had always represented a permanent, formalized declaration of our validity. The first Geo feature that was done on the early *Atocha* search had immortalized the original crew. With that in mind, EVERYONE now wanted to be captured on film, a mindset that sometimes produced some embarrassingly lame behavior in several individuals.

Unlike past experiences with National Geographic film crews, this new format used a director. Instead of passively filming us while we worked, the director "arranged" our activities somewhat to his liking. Although we sometimes felt that we were more actors than participants, none complained.

Ted was using his own personal boat to help charter people out to the site as well as drop off equipment and supplies. Sometimes Leah would sneak away from the office to join him. Since we rarely returned to port now, it was about the only time KT and I had to visit or gossip with them. It was also a great treat to show them the artifacts we found.

After the coin chest area of the ballast mound had been completely excavated, the exposed copper ingots scattered over the northern half of the feature were mapped and recovered. I was overwhelmed with notes, maps and coordinates pouring into the *Swordfish's* wheelhouse from divers on all three boats. When not diving, my whole day was spent working over the plotting table. We had barely scratched the surface of the ballast pile and already thousands of artifacts had been located.

Photographer Pat Clyne, and several assistants arrived from Key West with a structure they had built to photograph the ballast mound. The suspended camera would take pictures incrementally as it was moved over the feature. They hoped to capture the entire scene in a photo-mosaic as the Ribicoff Pegasus had done on the **Margarita** Main Pile. All the recovery boats were equipped with Nikonos camera systems that we used to document the individual artifacts, but in the poor visibility no grand photographic overview of the site had been attempted. At this point, the only representation of the entire **Main Pile** was the chart in the *Swordfish's* wheelhouse.

The recovery boats suspended operations to allow Pat, Ralph Budd and their helpers to photograph the site. Sadly, the time and effort invested in the project didn't return the expected results. Technical problems spoiled the mosaic and "my" chart would become even more important as the only accurate record of the entire **Main Pile.**

When our operations resumed, the raft of three recovery boats gradually shifted to the northern edges of the ballast mound as more exposed cop-

Silver bars began to pile up in the museum as the boats continued the recovery. Eventually 47 tons of *Atocha* silver would be on display. Photo: KT Budde-Jones

per ingots were recovered. Here the profile of the ballast stone began to recede into the mud. We communally used our blowers to gently dust away the lighter silt before proceeding with the airlifts.

The silver bar/coin chest area we had already excavated was only a fraction of the visible artifact field. There was still no sign of other standout items such as the missing eight bronze cannons or the registered gold bul-

A National Geographic cameraman films treasure trove coming to the surface.

Many of the artifacts found were personal possessions of the ill-fated passengers and crew. Two religious medallions and a lignum vitae cross were found amongst the ballast stones.

lion listed on the manifest. Perhaps they were in the three-foot deep jumble of mud, ballast and artifacts in the untouched portion of the **Main Pile**. We began to use hand held metal detectors again as we swam over the northwest extremes of the ballast, always on the lookout for large metal concentrations.

Little had been found of the missing gold bullion beyond the two bars Mike Meyer had found laying exposed the day the **Main Pile** was found. So far only a few gold coins and a small gold crucifix had been found; personal items from a wealthy passenger. Without referencing the manifest, we still didn't know for certain whether the bars Mike had found were actually the ones officially recorded as cargo, or the more common, unrecorded private "carry ons" of some rich passenger or businessman.

One afternoon diver Bill Moore was detecting and airlifting amongst a chaotic scatter of encrusted barrel hoops. He was working a short distance to the northwest of the ballast mound. Amongst the hoops suddenly appeared a new shape. As the encapsulating mud was gradually removed, seventy-seven gold bars were eventually revealed.

Gold items recovered in the past on both *Atocha* and *Margarita* site usually shone as brightly as new, despite their long saltwater immersion. These bars were different; their gleam was hidden under a thin frosting of calcareous concretion. Divers from all three boats got the chance to recover some of the cache, but most were brought aboard the *Dauntless*. Not far from

Dauntless divers examine some of Bill Moore's finds.

this spot a concentration of gold coins had been found as well. This developing area would soon have its own nickname, but it would have nothing to do with gold.

While examining the bars on deck we noticed they each had a unique marking that had never been seen before. There was an "X" scratched on the surface of each bar, looking as if it had been rendered with the cutting edge of a sharp knife. Was this a tally mark inscribed as each bar was listed on the ship's manifest? The artifact density and mud depth in this adjacent northwest area were much less than on the ballast mound. Kane kept working the *Dauntless* through this spot while the *Swordfish* returned to the **Main Pile** to continue our excavations there.

We added another dimension to our mapping process of the ballast mound. On the original chart I had decided to map all artifacts except the ballast stones. Drawing in the tons of ballast intermingled with the artifacts would have been too confusing. The chart now consisted of four overlays, each showing items found within a particular vertical strata. Now we would add another view - a sideways look at incremental slices of the ballast mound that would include the stones as well. It was slow going but produced a more rounded vision of the **Main Pile**.

True to the rumor, Mel had indeed bought the *J.B. Magruder*. For now it was being employed as a work platform for the freshly arrived "contract" archaeologists. Its former owner, Donnie Jonas, had been retained as captain and he had quickly assembled a crew to help him run the ship. The first job with which the *Magruder* was tasked was a formidable ballast study. They began by recovering tons of the stones that we had placed aside in the ballast dumps. The rocks were cataloged according to mineral content, weight, and whether they were quarried or simple river rocks. The idea was to determine the geographic origins of the ballast.

Very early one morning the subtropical dawn presented me with some real perspective on the *Atocha* and our search for it. The calm sea mirrored pastel clouds, while the unfiltered horizontal sunlight burnished anything protruding above the ocean's surface. Standing atop the *Swordfish's* wheelhouse with my back to the sun, I noticed how clearly our spar buoys were defined in the morning light. From my vantage point above the **Main Pile**, they progressed westward then turned northwest, marking the finds that had led us here.

The **Swivel Gun, Memorial Day, Nine Bronze Cannons,** the **Bank of Spain, Galleon Anchor** and the **Emerald Cross** spar buoys were all easily visible even though the latter ones were over seven miles away. With binoculars I could see the continued northward progression to **Fay's Cannon** and the **North Anchors**. The storm that sank the *Atocha* had deposited it on the bottom beneath our keel. The second hurricane had scattered the wreck over twelve miles. The spars testified to the impossibly long path.

Our activities on the **Main Pile** had become like shift work on a construction site. Radio transmissions annunciated via loudspeakers blared out over the nonstop diesel engines and hammering compressors. Everyone's time underwater was rigidly limited by the no-decompression dive tables we used. The first dive of the day was nearly one hour, which was followed by a mandatory two-hour surface interval. For the rest of the day all subsequent dives were limited to thirty minutes, followed again by at least two hours on the surface.

One diver might be working with an airlift, placing ballast rocks one-by-one into a removal bucket. By the time an artifact was found, he might be out of bottom time. The next diver would measure in the artifact, tag it, and record the data before having to surface. His replacement would photograph the object, and make more notes before recovering it. Thousands of artifacts were recovered in this tag team fashion. Crews on the surface kept track of the bottom times of each diver, and using steel hammers banged against the hull to audibly recall divers when their time was up. Each boat used a different hammer code.

I used our current work routine to spring a unique surprise on KT. During a brief reprovision turn-around trip to Key West, I covertly purchased a gold necklace for her upcoming birthday, an event that would happen while we were at sea. When the big day arrived, I tied an artifact tag to it emblazoned with "Happy Birthday" and snuck it to the bottom during my first dive of the day. She was already working on the other side of our airlift excavations when I reached the sea floor.

KT was so completely absorbed in her own activities that it was easy for me to bury the chain in the mud on the edge of the excavation trench. I left just the smallest part of the necklace visible and then swam over to her, indicating that I had just found something interesting. She followed me and immediately spotted the bit of chain. She half seriously pushed me aside and hand-fanned away the mud around it. Imagine her surprise to find the gold

The face of the excavation into the *Atocha* ballast pile

chain already tagged with a birthday greeting on it!

The *Swordfish* crew continued to systematically work northward through the ballast mound, uncovering more of the wooden structure as the dig progressed. The lower portions of at least seven of the ship's massive ribs were found. These huge structures were not fashioned from one piece, but were originally constructed in segments. The lowest, central part of the rib was called the "floor". The pieces attached to the floors ,which together, fashioned the curve from the bottom of the ship to the sides were called "futtocks".

The massive floors and futtocks rib components found were all skewed in the same direction like toppled dominoes, suggesting that the ship was still moving forward with some speed when it made its final impact with the bottom. There were a large number of outer hull planks under and around the ribs with a blatant gap right where the keel and keelsons should have been. Had this missing portion of the keel been ripped off on impact with the outer reef? Another surprise was a length of fragile hemp rope found buried under the timbers.

One September afternoon we had just located an intact ceramic jar miraculously undamaged by a crush of copper ingots when we got a radio call from the *Dauntless*. The divers had made an exciting discovery while cleaning up the already airlifted gold area with their mailboxes. Mixed in with the crumbled seashells and calcareous fragments had been found several uncut emeralds.

Once the *Dauntless* divers had tuned in to the shape, size and color of these latest finds, more were recovered. Mel was notified on the radio about the development and he was absolutely ecstatic. In short order this new flavor of artifact gave the gold area its permanent name - **Emerald City.**

With his father's obvious blessing, Kane focused all of the *Dauntless'* attention on the spot. As more loose, uncut emeralds were found, the adrenaline frenzy increased but little care was needed here as all the other artifacts had already been mapped and removed from the spot. The blowers could be used indiscriminately, rechurning through the shelly seafloor remnants without the risk of disturbing anything else. There was no way other than with the diver's own eyes to find the stones. To present the divers fresh opportunities, the *Dauntless'* mailboxes blew the bottom detritus again and again - much like reshuffling a deck of cards.

This latest situation worked out really well for all concerned. The *Swordfish* could continue to very carefully work and map the ballast pile. Kane happily recovered valuables at his own pace with few academic restrictions. *Virgalona's* crew filled in where needed. While airlifting on the westernmost edge of the ballast pile, I found a good-sized emerald still embedded in the mud. Others turned up as well. The *Swordfish* crew decided to keep those particular finds quiet to maintain the happy equilibrium.

No one is able to function at one hundred percent while at sea, even under the best conditions. All the boat crews were now working on limited reserves. The silly season when burnout began affecting everyone was well upon us.

Since this year's operations had been going hard since early spring, our equipment and people were no longer finely tuned. With no time for maintenance or repairs, hard use and the saltwater environment significantly degraded the boat's appearance and equipment. The quality of life aboard had slipped a few more notches. At least the heightened workload absorbed most of the extra energy needed for pursuing interpersonal disputes. The crews numbly coped with the additional stress of visitors being unloaded on us almost daily.

With tired eyes the *Swordfish* crew finally finished the ballast pile-mapping project. We had already worked through the bulk of the known **Main Pile**, but many significant manifested artifacts remained missing. More registered gold bars, eight bronze cannons, at least two hundred silver bars and many thousands of coins were still at large. Though spectacular, it was clear now that the **Main Pile** was not the repository for all of the *Atocha's* missing artifacts. Where were all the missing items? We had all thought that once the **Main Pile** had been found that the *Atocha's* story would be completed. Our treasure hunt wasn't finished after all.

While looking for the **Main Pile**, we had skipped lightly down the trail of secondary scatter that lead us to it. The vast majority of the area between our secondary scatter finds remained unexplored. Tom Ford decided a change of pace was needed. He started careful excavations to the northwest of the ballast pile, working the trail of artifacts back in the direction from which we had originally come.

We picked up the track not far from where the *Dauntless* had first intercepted it, following the path away from the **Main Pile** one excavation at a time. We were now moving in the same direction that the second storm had driven the displaced upper hull of the *Atocha*. The artifact strata was about two feet down in the mud, so we dug with our blowers almost down to where the strata lay and then circle searched the excavation with metal detectors. The fifty foot wide track of copper ingots, barrel hoops, coins, ballast and spikes lead strongly to the northwest for some distance. At just over five hundred meters distance from the **Main Pile** we found a six-foot long grapnel hook. From this point the artifact track narrowed to about twenty-five feet and changed direction, now moving west-northwest.

Was the grapnel left by Melian or the other original salvors prospecting for the *Atocha's* remains? Possibly, but perhaps it simply had just been aboard the *Atocha* itself. Whatever the case, the path of artifacts didn't point at all to the **Swivel Gun**. When the trail turned to the west-northwest, it lined up perfectly with the distant **Memorial Day** and **Bronze Cannons** spar.

In time we came to understand the significance of the **Swivel Gun** feature. It was an isolated grouping of small, heavy artifacts with no direct correlation to the artifact trail passing well to the north of it. Dr. Lyon's research gave us a likely reason for its existence.

Not long after the massive second storm had stirred through the area, the Governor of Havana had decreed that buoys should be maintained permanently near the assumed spot of the *Atocha's* demise. It would help to guide their future salvage efforts.

The Spanish had used anything weighty to anchor their buoys. Chain shot or even an old swivel gun were convenient, much in the same way we used concrete blocks. Our styrofoam buoy balls and PVC spars were just modern versions of their small floating barrels and wooden spars. Vargas and Melian had been here. They had been close, but not close enough to make a difference. How many times had we been in the same situation?

This sailor's pipe bowl still contained ash from the last time it was smoked aboard the *Atocha*.

Coda

Tom Ford had personal plans that took him away from Key West. I inherited command of the *Swordfish* in his absence. It felt very natural and a little bittersweet to once again be in command of the small ship's crew and its destiny. I was determined not to waste the opportunity.

The *Magruder* had finished her ballast study, and her fresh-faced team of contract archaeologists had made another airlift sweep through redeposited silt in the area where the silver bars and coin chests had been located. Now they and the *Dauntless* were tearing up **Emerald City** like two hyenas on a kill. Emerald fever had swept the company and no trip was considered successful without finding a handful.

I found looking for emeralds mildly entertaining, but with the two other boats hunched over the best pickings there wasn't much room left for *Swordfish*. On a whim, I positioned our boat about seventy feet further south of the southernmost **Atocha** **Main Pile** area and we began prospecting with airlifts.

The mud was noticeably deeper and our excavations revealed a surprisingly strong deposit of ballast and even more ship's timbers. Over the next few days we slowly exposed another intact floor timber and lengths of outer hull planking. How did this material relate to the **Main Pile**? Its meaning was vague but intriguing, like a freshly broken twig found on a long abandoned pathway. I had experienced this same feeling years ago as a freshly minted captain working the *Margarita*.

At the end of our trip, Ted summoned both KT and me for a luncheon meeting with Mel. After returning from a trip of his own through the Caribbean, Mel had special plans for us.

The luncheon subject wasn't what we expected. To our horror we learned that Mel intended to take us both off the *Atocha* project and have us set up a survey for shipwreck sites on some little known island near Puerto Rico. We tried every excuse we could think of to get out of it, but Mel was adamant. I repeatedly brought up the importance of continuing the site mapping of the *Atocha*, but Mel deflected that by saying that the data could still be collected and that I could plot it when we returned.

As Mel saw it, KT and I were experienced with magnetometers and other survey gear, we could set up a search program and were reliable and resourceful enough to accomplish what needed to be done with no supervision. He tried to lessen the sting of being taken off the *Atocha* project by adding, "You guys go down there and get things all set up and started. I'll send somebody down in about a week to relieve ya."

We didn't know it yet, but Mel himself had psychologically moved on from the *Atocha*. To him, the discovery of the **Main Pile** closed the book on his sixteen-year search, leaving a sudden void in his life. Though he certainly

This gold crucifix with two pearls and two gold coins were found in the same area as the uncut emeralds.

enjoyed the publicity of continuing *Atocha* finds, the inertia of his company's discovery led him to quickly seek out his next BIG THING. With more investors willing to throw their support and money at him, Mel's entrepreneur status had gone blue chip.

Mel's great success on both the *1622* and *1715* fleet sites was owed in large part to archival research done by Dr. Lyon, Kip Wagner and others. This information produced specific targets of real ships and known cargos. Now, his once exceptional accomplishments on the *1715* fleet sites and *Santa Margarita* site paled compared to the against-all-odds finding of the near mythical *Atocha*. Mel had finally achieved what he always wanted; the uncontested title of "The World's Greatest Treasure Hunter". Giddy with self-confidence and optimism, he naturally reverted to his favorite technique of all…wildcatting.

This time wildcatting would be on a much grander scale. Instead of randomly digging holes on the ocean floor looking for a single ship, he would personally wildcat on an international scale. His interest in the use of archival research had faded; it was obvious to him by now that he could probably find treasure anywhere.

While on a freewheeling recon-trip through the Caribbean, Mel had accidentally ended up on the small island of Vieques. In his wanderings he soon met two people who would inspire a whole new venue. One was a Puerto Rican businessman struggling to turn an old sugar mill ruin into a tourist resort. The other, a local fisherman, told Mel that he too had once found a small gold bar while diving in nearshore waters. Of course, the fisherman had no photographs of the item, and had supposedly sold it to someone whose name and address he did not know. That didn't matter at all to Mel, in short order he developed a plan.

Mel and the businessman would become partners in the resort. Mel would quickly set up an electronic survey to find treasure-laden shipwrecks in the surrounding waters. Visitors would flock to the resort in great numbers, drawn by the opportunity to join in on the underwater recoveries. The guests would fill the resort's coffers, and many would become investors in the underwater projects. They would also serve as free labor on the dive sites. Mel would build a local museum to house the finds, which would bring in even more visitors and revenue. What could possibly go wrong?

One minor detail that Mel overlooked was that the routes the treasure galleons took were nowhere near Vieques. There was also no archival evidence that any notable ship had ever sunk in these waters. Mel easily brushed that off. Hadn't the fisherman found a gold bar? Did he not also say that he had seen old anchors and a cannon lying on the bottom? That was more than enough evidence for Mel.

KT, myself and *Swordfish* diver Paul Busch were sent down to immediately start up the survey in October of 1985. We were to live in the mostly unfinished sugar mill resort and use the fisherman's boat. Mel and his lawyers had already met with several Puerto Rican government officials about his plans, and was now drumming up investors.

In the long run, even Mel's overheated imagination and limitless optimism couldn't overcome Puerto Rico's third world political genetics. Our examination of the local waters turned up nothing more than a few lost anchors and a small iron cannon that someone had tossed in the water to use as a mooring. We broadened our search while Puerto Rico prepared for its upcoming electoral cycle.

Our presence on Vieques suddenly became a hot potato between the three battling political parties, partially inflamed by Mel's earlier, island-wide bragging that we would be finding "turtleshells filled with emeralds". The locals watched us returning to the docks every afternoon with nothing other than our survey and dive gear, so to them we were obviously stealing and hiding the vast treasure (the poverty-ridden Puerto Rican people's birthright, etc.).

All this quickly degenerated into arrest warrants, Puerto Rican Senate hearings, fist shaking politicos and bureaucratic denials. Our innocent one-week survey trip turned into three months of good old fashion banana republic hide-in-the-jungle political adventure. Not long after the 1986 New Year, the whole project finally imploded. Mel telephoned the order to "Get outta there!" Paul, KT and I gladly escaped from the Hispanic histrionics back to Florida.

It was great to get back to Key West again. Just before we had left for Vieques, KT and I had moved from the Miguel's "mother-in-law" apartment at 417 Cactus Dr. to a studio apartment of our own in New Town. Always looking out for our interests, Ted and Leah dutifully made sure the rent was paid during our unexpectedly long absence. We took some time to settle into the tiny apartment, and it quickly felt like home.

Our relaxation didn't last long. Tom Ford needed a break running the *Swordfish*, so I agreed to captain it again in his absence. Before leaving for Vieques, I had told Tom about the artifacts we had found just south of the **Main Pile** but he hadn't acted on my tip. No matter, it gave us something on which to focus on this trip. KT, Paul Busch and I happily joined the regular standing crew of Mike Moore and Christian Swanson.

The other boats were still busy working over **Emerald City,** and gave no notice to the *Swordfish* positioning itself some distance away. I set us up about 400 feet to the south-southeast of the *Atocha* **Main Pile**. Since mud was deepening, we cleared off the upper two feet with our blowers before reverting to exploration with airlifts and detectors.

Our probe of this new area almost immediately showed promise. We found two intact "olive jars" and large fragments from several others.

Coda

Additional large planking timbers were found as we extended our survey benchmarks from the **Atocha Main Pile** directly to this new spot. This latest artifact trail appeared to project about 165 degrees from the **Main Pile**. As the trip ended, all the *Swordfish* crew agreed to downplay our finds to the rest of the boats. This development was our discovery, and we wanted to have the chance to exploit it.

Tom Ford returned in time for our next trip, and for a while the *Swordfish* participated with the other boats in an **Atocha Main Pile** ship's timber study and recovery organized by Duncan Mathewson. When that project was completed, *Dauntless* and *Magruder* immediately jumped back into **Emerald City**. Tom Ford listened with growing interest to my southeast discovery stories. Excitement spread throughout the crew when he decided further investigation was in order.

To my surprise, he didn't set up the boat in the area where I had left off, but jumped yet another hundred and fifty feet further southeast. We dusted briefly with our blowers to remove the loose silt and began to search with detectors and airlifts.

The detectors began to ring out almost immediately. A lightly sulphided silver plate turned up followed by a really remarkable find - a very large and completely intact ceramic jar lying on its side just under the surface of the mud. It took hours of careful excavating to expose the jar while other *Swordfish* divers explored nearby. Cooking pots, gilded silver cups and hundreds of loose coins were located.

The next day I got several detector hits in close proximity. Alone in the dim gray-green with my airlift, I gently excavated the first, partially exposing what looked like another large, intact ceramic jar. Before I ran out of bottom time, I decided to dig out the other nearby target. The airlift exposed a small wooden box, about eight inches square. As I quickly mapped in the box I noticed a small, light colored emerald lying next to it in the mud. The digging frenzy going on in **Emerald City** would be transposed here in a second if the word got out to the other boats about this. All I could think of was "Oh shit."

This new scatter was as rich in artifact diversity as the **Atocha Main Pile** had been. We found keys, locks, silverware, hairpins, spikes, knives, gold money chain, silver bowls and many more coins. My weedy little trail had become a major artifact highway. The *Swordfish* crew wanted to keep the area for our own exploration, but at the end of our trip Tom decided to tell the other boat captains about what we had been finding and where. This initially caused some hard feelings, although the word would have gotten out anyway when we returned to Key West with our finds.

When we arrived back in port, the little wooden box I had found was x-rayed to determine its contents. It was so densely packed with metal that no clear image could be seen.

Always sensitive to a great publicity opportunity, Mel's office staff made arrangements for the box to be opened live on "Good Morning America." Actor Cliff Robertson (who was going to play Mel in an upcoming TV docu-drama about Mel and the Fisher family) was front and center, bracketed by contract archaeologist John Dorwin and Mel. The three of them sat behind a table that displayed the box for the TV cameras. Other hangers-on quickly crowded around, pushing the *Swordfish* crew so far behind that we appeared to be onlookers.

When the moment came for the box to be opened it didn't cooperate,

A large ceramic jar found on the **165 degree line** is gently brought to the surface.

Coda

The wooden box   Photo: KT Budde-Jones

From left: KT, myself, Cliff Robertson and Mel Fisher wait for the opening of the wooden box.

despite having been previously prepared. During the next commercial break it was "massaged" again by a rather wired conservator.

The second opening attempt was more TV friendly. On cue, the lid was removed and the contents gently removed and displayed for the camera.

Nearly fifty feet of gold money chain had been packed in the box along with several stacks of silver coins. I had thought it had felt overly heavy, even as I first lifted it from the mud.

Predictably, the emerald area was abandoned and all the boats moved down to the newly christened **165-degree line.** The *Swordfish* had already extended the benchmarks from the ***Atocha* Main Pile** along this new trail at

Recovered stacks of encrusted silver plates, candlesticks, and silverware after being put in an onboard storage tank

Coda

A silver pitcher being excavated Photo: Tom Ford   The same pitcher (below) with other recovered artifacts after being stabilized in the conservation lab   Photo: KT Budde-Jones

thirty-foot increments. The other boats began to contribute data as more finds were made. I instructed contract archaeologist Corey Malcom on our procedure so he could bring the others up to speed on the *Magruder*. Everyone was finding wonderful things.

We found and mapped impact scars where the *Atocha*, in its sinking condition, had actually slammed into the dense mud as it settled ever deeper driven by the hurricane's waves. Fragments of wood, which sometimes included spike heads sheared off by the force, were smashed into the scars An intact sea chest apparently owned by Pilot (navigator) Martin Jimenez was found not only containing his personal wealth but a number of navigational instruments as well. Individual astrolabes were found along the trail and even the ship's bell, broken into pieces was recovered. Silverware, more keys and locks, pottery and galley ware were found in abundance. Not far from the pilot's chest location, an accidental find would tickle Mel's fancy.

A *Magruder* diver had set his airlift aside while working on an artifact recovery. He violated the operations policy against leaving an unattended airlift running, and it began digging down into the mud on its own. The tide was nearly slack, and anything that got sucked up the tube tumbled out of the upper end and fell almost straight down. Our hero suddenly noticed green crystals drifting down all around him. It was raining emeralds.

The airlift had dug down into a pocket of the stones and was now spewing them out. They may originally have been kept in a cloth bag of some sort, although no fragment of it was ever located. Soon everyone was picking up loose emeralds, the stones rapidly filled a large Mason jar.

These stones were very light in complexion and were not the same high quality as those from the **Emerald City** area. Mel didn't care, they were still emeralds and they would bring in more investors and publicity. For a while there was a repeat of the **Emerald City** frenzy to find more, but this feature was far more localized and played out quickly. By comparison to the other emerald finds to the north, these lower grade stones looked more like green tinted quartz. They were sometimes referred to as "fish tank gravel" by the more cynical crewmembers.

The **165 degree line** work progressed further away from the *Atocha* **Main Pile.** Eight hundred feet out, the trail just faded away. We now knew the direction the ship was pushed by the storm as it was sinking; a path indicated by the artifacts it had dumped right up to the **Main Pile** location. Of course, the 165-degree heading was just our observed compass bearing along the artifact trail as we worked away from the **Main Pile**. The actual direction that the ship would have traveled towards the **Main Pile** position was the reciprocal, or 345 degrees. It would be a future project to extend the line out to the distant outer reef shoals in an attempt to determine the place where the *Atocha's* fatal impact hull damage had occurred.

The *Saba Rock* was back on the site, this time being captained by Don

Durant. After months of inactivity, a contract was finally reached with Mel that allowed them to again work the wreck. They were not allowed on the ***Atocha* Main Pile**, so they spent their time working along the scatter trail up near where the **Memorial Day** emerald jewelry find had occurred. This time the promise of the jewelry deposit wasn't squandered. They were at last able to relocate the spot, and Treasure Salvors diver Christian Swanson helped recover another 62 pieces of the same jewelry that KT and I had first found.

As *Dauntless* and *Magruder* shifted back to the **Emerald City** area, *Swordfish* continued exploring the trail of secondary scatter leading to the west-northwest. We started about 200 meters from the ***Atocha* Main Pile** and began working along the largely unexplored track. Coins, copper ingots, ballast and lots of barrel hoops were located. Could the missing cannons and other unfound manifested cargo be between here and **Memorial Day**? A number of days were also spent around the **Swivel Gun**, which only confirmed that it was not part of the wreck, but a buoy site established in the 17th century.

We were all spending so much time at sea there was even less direct interchange between the office staff and the boat crews. At the end of each trip we would quickly unload our recovered artifacts and begin the quick reprovisioning and refueling necessary to get right back out to the site. The shore-bound staff had gotten quite large with lots of new faces, their occupations mostly unknown to us. Archaeologist Duncan Mathewson was already publishing his newly written book on the find, which explained why we hadn't seen much of him out on the site. As each boat cycled back to port for a day or two, it bought back more gossip to be circulated amongst the other boats.

From 40 miles out to sea it was hard to tell truth from fiction in these stories. "A TV movie was being made about Mel, Deo and the *Atocha* find." "Beer companies were going to use the divers in their commercials." "The company is going to buy a submarine." "Gene Lyon is updating his earlier 1979 book on the *Atocha* to include all the latest events." "The office staff all make more money than the divers." "The company was going to be liquidated." We could never tell which one of these stories was true, even news about internationally important events like the Chernobyl disaster rarely made it out to us.

The favorite topic of speculation was the upcoming division of the *Atocha* material found in 1985. Even after the *Margarita's* rich recovery, the number of artifacts and investors involved in the *Atocha's* **Main Pile** year of discovery overwhelmed us. We had heard that an independent company was hired to develop a computer program that would fairly and randomly select how the items would be distributed. We all knew that the investors and the Fisher family would certainly get a percentage of the artifacts, but what about the boat crews? How did the Mel Fisher Maritime Heritage Society (the separate, non-profit organization that ran the museum) figure into all this? None

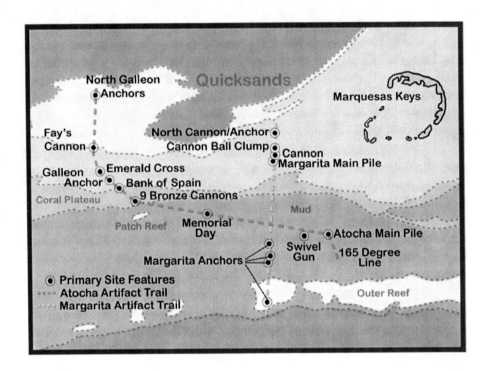

of us really knew.

Taffi Fisher had taken on the huge task of making sure that everything that was found in 1985 would be ready for the division. Many thousands of artifacts had to be cleaned, stabilized, recorded, studied, photographed, cataloged and certified. Each was evaluated by experts and given a point value relative to each other. The total of all these points would be what was available to divide.

Taffi and her staff had to figure out what each investor's percentage of the total points would be. The computer program would randomly select items from the points available and "fill" his percentage. If an investor was eligible to get 1000 points, he could possibly get one item worth a thousand points, or ten worth one hundred each. The computer program was designed to make sure that there was no inside track to improve one investor's chances over another.

None of the boat crews were really sure how they fit into Mel's plans. A few times over the years when talking about the *Atocha* (usually after a few rum and cokes), Mel would sometimes proclaim to the divers, "We're all goin' to be millionaires." He hadn't said that since the **Atocha** **Main Pile** was found. No one really believed we were going to get rich, but it was expected that we would at least get a little something out of it. The problem was that there was no specific amount officially set aside for each crew member, it was totally up

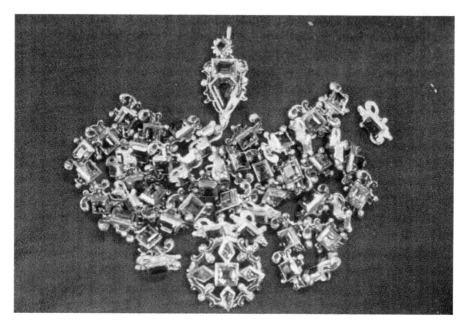

More jewelry found at the **Memorial Day** area   Photo: KT Budde-Jones

to Mel's generosity on that particular day. A nasty rumor floating around was that Mel felt the divers had gotten too much during the ***Margarita* Main Pile** division, and this time he wouldn't be so big-hearted.

Ted considered himself the chief advocate for the operations staff, and met with Mel over the matter. He learned that Mel indeed did have a plan for his employees and it was simple in the extreme. One percent of the total points had been put aside for all the staff, and that amount would be equally divided amongst them.

The company had recently swelled to about sixty employees, most of whom had been hired on after the ***Atocha* Main Pile** had been found. Mel's plan would give equal reward to those who had just been hired as office assistants as well as to crew members and boat captains who had endured years of hardship to realize the dream.

Ted had never been shy about disagreeing with Mel. This was a serious issue to him, and he was not prepared to accept a compromise. He wanted his operations people taken care of.

Ted's determination outflanked Mel's usual distractions, and the two remained in an undisturbed meeting until both were satisfied with the result. Ted may not have won a popularity contest with all the employees, but his negotiations brought the greatest gift to "his" crews.

New employees would now no longer share equally with the old hands. A minimum time of employment was agreed upon for staff to receive anything other than normal pay. Crewmembers would receive an amount based on length of service, position attained, personal motivation, and value to the operation. The icing on the cake was that Ted had even convinced Mel to kick more percentage into the operations employee's kitty.

Ted quickly formed a committee to evaluate all the boat crews so that their percentage of the bonus could be determined. Tom Ford, Andy Matrocci, Dick Klaudt, R.D. LeClaire and myself were tapped for the duty. We all walked away satisfied that the process had been fair and that Ted's "program" for the task had worked very well. It was a timely event, as the division would happen soon.

We learned that a TV movie on the *Atocha* had indeed been filmed. Mel's old friend Cliff Robertson played Mel and Loretta Switt was Deo. Details about the project hadn't filtered down to the crew, but we did hear that it had been filmed in the Caribbean. Of far more immediate interest to the divers was that Mel and Deo had decided to present Super Bowl-type rings to the employees involved with the **Atocha Main Pile**. At the first opportunity we all went in for ring sizing.

Several months before the scheduled CBS play date of "Dreams of Gold," there was a private screening for investors and friends of the company at the Tennessee Williams Theater in Key West. The Fisher family in particular seemed quite happy with the production and the investors were thrilled to be a part of a nationally recognized story.

The boat crews were too busy taking advantage of the late season's good working weather to attend the screening. We all waited until November for the officially scheduled broadcast on TV.

After seeing the November broadcast, the divers and captains were less than complimentary - the screenplay had completely ignored their existence. With the exception of the graphic *Northwind* sinking, most saw the whole thing as a bad, largely fictional insult. Numerous comments were made about it being "The Goddam Brady Bunch goes treasure hunting." The operations employees now hung their hopes on the yet-to-be-aired National Geographic special.

The magic date of October 13, 1986 finally arrived. At 2:00 in the afternoon, Taffi Fisher pushed the key on the computer that would begin the division program. There was media present to record the event, but after this opening gesture there was little to talk about until the computer finished its task. After almost a full day of working out the selection process, the computer spent three days printing out the results. All the employees regularly checked the printouts to see if theirs had come up yet.

KT and I had lumped our tiny percentages together to increase our chances. Near the end of the printout cycle, our division bonus finally accordioned out of the printer on green-barred paper.

The division printouts were officially distributed to the investors, crew, Fisher family and the Museum before the regathered media. This was a far friendlier story for the press than the earlier start of the computer run. On national news it was reported that many of the investors and divers were now millionaires. The TV footage showed office staff member Sherry Culpepper's emotional acknowledgement of her own printout.

Thanks to the generosity of Mel and Deo, the Museum had been granted a nice chunk of their personal percentage. Favorable tax laws encouraged the philanthropic leanings of several investors to do the same. Between artifact donations previously made by the Fisher family to the not-for-profit Museum and this latest gift, the Museum would permanently own most of the notable finds and a broad cross section of artifact types from both the *Atocha* and the *Margarita*. The Museum had now reached world-class status and the number of visitors only increased.

Most of the crews were all smiles with what they had gotten, but there was still lots of printout comparing. Ted and Leah had been well rewarded; among other things they had drawn an entire uncleaned coin chest, an unexpected and potentially valuable windfall.

One distinctly displeased employee was Mel's long time associate and electronics guru, Fay Feild. Fay stayed only long enough to receive his treasure and then permanently departed the Keys and Mel's employ. He had been one of the few unique characters who really defined Treasure Salvors as a company. For years he had adjusted the finicky magnetometer, fixed broken detectors and kept the Del Norte system maintained. We would have a much more difficult time without him.

Treasure Salvors Inc. rode off into the sunset. This rumor had come true, Treasure Salvors as a corporate entity was no more since tax issues required its dissolution. A new company, Salvors Inc. replaced it. To the employees this meant little, the only thing that changed as far as we could tell was the name on our paychecks. Another change was that the *Virgalona* had steamed away from Key West for good, a gift to Momo from the Fishers. Over the last year the tired old workboat had quietly been retired from the *Atocha* site; it had proven ineffectual in the deeper water of the **Main Pile**. Momo would put it to good use again in the shallow waters of the *1715 Fleet* sites.

Taffi and her staff shifted into the next phase, the physical distribution of the division. Investors and employees alike made appointments to come and pick up their portion. KT and I had to wait several weeks for our date.

When the big day finally arrived it took over an hour with Taffi to check off our allotted portion. The majority of our share was in silver coins.

We also received a few silver bars and a large uncut emerald. We had no idea what to do with the artifacts, so we rented some safety deposit boxes at a nearby bank. All of us were aware that 1986 had only a few more months left before its numerous finds would also be going through the same process. More safety deposit boxes might be in order.

In between the late fall weather fronts I was presented with an unexpected opportunity. The *J.B. Magruder* was short a captain for its next trip. Since Donnie Jonas wouldn't be available to take it out, Mel asked me to do so. I had always been intrigued by the big boat's potential, but had never spent any time aboard it. I was happy to take the helm.

By now the contract archaeologists had departed and the boat was down to its standing crew. I made no attempt to manage them; they knew their own routines and the ship's operations better than I did. My job was to get the boat out on the site, conduct safe, useful operations and bring it and the crew home in one piece. When not underway or otherwise engaged, I just quietly watched the crew dynamics. They did what I asked, but in their eyes I was definitely just a temp.

I was surprised to learn how slow everything happened on the big boat. Maneuvering, setting up and anchor recovery required much more forethought, time and effort than the much more lithe *Swordfish*. The *Magruder* was vastly more comfortable to live aboard with lots of room, almost limitless

Taffi Fisher, KT and I go through the division process.   Photo: Leah Miguel

fresh water, air-conditioning and vast fuel tanks. It was also much more complicated.

A few more trips to the site on the *Swordfish* finished out the year. There was more talk among the crew regarding the latest rumor; Mel was selling the company. With no evidence, it was just another time-killing conversation of little importance.

KT and I planned on visiting family in Wisconsin over the holidays. Friends who owned a nearby scuba diving shop suggested we bring some of our treasure. They would put on a display in their store, it might bring in more potential customers for them and we might be able to sell a few coins. It sounded like a good idea and it would be fun to show off a little.

The dive shop display would turn out to be wildly popular. Even with the modest media contacts that had been arranged, thousands of people crowded into the store to see the artifacts as well as meet us. After all our isolated time at sea this was our first taste of how big the *Atocha's* story had become. Even Menomonee Falls, Wisconsin knew of it. We also sold enough coins to know that we were on to something.

Numerous marketing plans for the investor's treasure items were already in the works. The Fisher family would no doubt also be trying to turn their portion into money as well. KT and I knew that if we were going to turn our bonus into dollars that we would have to get a jump-start on the rest. This just might be the way to do it, being sponsored by dive shops in affluent cities. The dive shops would get a new audience for their dive instruction and we could sell some of our coins. All the years of low wages might finally be offset if we did this right.

May You Live in Interesting Times

1987 started with new gravities pulling in several directions. Last year's scuttlebutt about the company being sold had turned out to be true. This was hard to imagine for those of us in the trenches. For as long as we had known Mel, he and the quest for the *Atocha* had been two sides of the same coin. Our emerging reality was that Mel had moved on, and that the expedition had not only become a business, but a commodity.

For many of the crewmembers, it was curious why someone would want to buy the *Atocha* and *Margarita* sites after both had been so extensively worked. We all knew there was still unaccounted manifested cargo and a likelihood of more unmanifested surprises such as emeralds, but the general feeling was that we had already found the bulk of the missing cargo.

The Museum had become the biggest benefactor of the company's sale. In purchasing Treasure Salvors and its assets, the new management now owned the company's percentage of previously found *Atocha* and *Margarita* artifacts. They quickly donated all these premier items to the non-profit Museum, assuring the permanency of the world-class collection. Those of us excluded from the wheelings and dealings of the office cynically guessed this gesture had more to do with tax write-offs than generosity, but it did keep our keystone finds from being scattered.

From the boat crew's perspective, the intentions of the new Georgia based owners were hard to decipher. The only changes we noticed was another new company name on our paychecks; Jamestown Treasure Salvors. As far as I could tell it was still business as usual.

While trying to digest this latest serving, Tom Ford added a little more salsa. He had heard that Donnie Jonas had tired of running the *Magruder* and that I was being considered to captain it. This made no sense to me since Dick Klaudt had been acting as Donnie's first mate and was more than qualified to run the boat. I confidently discounted the whole story as just another example of more "T.S.- B.S."

Reliable shore-bound employees began telling us how the new owners would sometimes make unscheduled visits to the office, and had already developed relationships with a few opportunistic members of the operations staff. A few of the *Dauntless* divers and ex-*Virgalona* crew were getting cozy with the new bosses, hoping to gain favor in this evolving regime. Some even bragged about having the principal partner's personal phone number. I didn't even know his name. KT and I had a phone put in our little apartment so that our office friends could notify us of the next visit while we were in port.

Within a week I got a call from Ted at our apartment. "Donnie's not interested in running the *Magruder* anymore, and I'm tapping you. Interested?"

"Well...I dunno...maybe. But what about Dick? Why isn't he taking it? Another thing, remember that treasure marketing idea that KT and I had talked to you about? We're going to need some time off now and then to do that. We're already booking locations to do it over the next few months so I won't always be around. It's going to be hard to be a boat captain if we're having to take off a few days at a time for a while."

"Dick says he isn't interested in running the boat himself... he wants you. I'll talk to him and have him give you a call. You guys work it out."

Dick called the next day. As Ted had mentioned, he just didn't want the responsibility completely on his shoulders. I explained that KT and I would be taking off for long weekends several times a month for the next few months to try and market our treasure. He had no problems with that, and would run things in my absence. During the conversation we discussed a dual captain concept that seemed to work for both of us.

As our talk continued, we both agreed that there would have to be some personnel changes made. Some of the existing crew were problematic and would only get worse with this change in management. Right away it sounded like Dick and I were in lockstep. We already respected each other's abilities.

My first official day as "co-captain" on the *Magruder* was dramatically different from my trip aboard the previous year. This time there was an undisguised distaste for my presence by a few of the ship's "old hands." None of them really knew much about me other than I was close with Ted, but that was enough. Vince Trotta was brave enough to vocalize what some of the others were thinking. "Man, why are you here?"

I smiled. "I'm here just to piss you off."

At this point in my career I was miles past the point of being intimidated by crewmembers. I worked with those who had potential. Permanent malcontents had to go. Dick and I had an agreement that neither of us would sack a diver without first conferring with the other. We had already agreed to a short list of those who might have to be discharged soon. I decided to rock the boat first.

There was one individual on board who had not been working out for some time, and there were many complaints about him. I terminated his employment as painlessly as I could and hoped that this quick, decisive action would impress upon the others that there was a new sheriff in town.

It worked for a while. Dick and I had big plans for the boat and we were going to need as many people as we could keep. The *Magruder* had sat at the dock for some time until Donnie sold it to Mel. During that time it had received only the most basic maintenance. Using it nonstop for the last year had made it even more raggedy. It was an old boat with lots of hull and bulkhead rust issues, and some of the machinery was worn out. It was time to haul the boat out of the water for a major repair and refitting. This kind of unsavory work requires lots of busy hands.

There were no boat yards in the Keys that could handle the *Magruder's* size, and going to a commercial yard upstate would require that we hired the yard's employees to do the work. The expense of this would be horrendous - especially since our own people could cope with the project. We looked for another answer.

There was an old boat yard on nearby Stock Island that was currently in a legal limbo. The overseer of the property agreed to let us use the yard but there was no lifting equipment on site. Steve Candella of Key West Welding stepped up to the challenge (as he had with so many of our other unusual projects), and together we came up with a plan. There was a single slip in the middle of the yard that was just big enough for the *Magruder* to back into. We would lift the *Magruder* out of the water using seven cranes and then build a steel framework suspended across the slip under the ship's hull. The *Magruder* would then be lowered down onto the supporting steel structure and secured. Wooden work floors would be built around the steelwork and no one who visited the site would guess that the *Magruder* and its workers were actually suspended over water.

Dick and I began expanding our crew roster with an emphasis on

Rebuilding the *Magruder* in the boat yard... the large back deck would prove to be an excellent work platform.

hiring divers who were skilled welders. When the day came for me to back the *Magruder* into the boatyard slip, our guys and Key West Welding made short work of it. In one day's work we had the *Magruder* on its suspended steel cradle. Now the long, dirty process could really begin.

Just about this time the company employees got a 10.0 Richter shock in the form of a W-2 tax form. Unbeknownst to us, Treasure Salvors' tax attorneys had decided that any artifacts given to the employees was a bonus that would require taxes for the value of those items to be paid by the company stockholders. The majority stockholders were the Fisher family.

The only way to relieve the stockholders of this burden was to declare the bonus on the W-2. Pieces-of-eight don't compute on a tax form, so the bonus was listed as earned income. Several formulas including Mel's famous "Fisher Factor" had been used to calculate a monetary value to the items and the result was sky-high, way beyond what the average person could sell them for.

The Federal Government didn't care about the treasure, it had W-2's that now showed that we had all received extremely generous bonuses and it was time to pay the piper. Oh yeah, and this was the last year that the piper was eligible to get 50% of the total. Between KT and I we owed tax on many hundreds of thousands of dollars. Pretty impressive territory for people who actually made several hundred dollars a week.

This new revelation went through the staff like barracuda through a baitfish school. Disbelief, hand wringing and muttered threats were the universal reaction. Everyone expected to have to pay tax on the material IF it was sold, not before. Some had planned to never sell any of it. The foundational Treasure Salvors did offer to pay the taxes for any employee who wanted to return a portion of or all of their treasure, but it would only credit a fraction of the W-2 declared value of the items.

Several meetings were quickly set up with expensive tax attorneys who, sensing blood in the water, wanted to represent the divers as a group. Auction houses were considered as a way to get quick cash, but no one was sure what kind of prices that venue would bring or how quickly. The clock was ticking and tax time was coming up.

Mel offered to pay for a few hours with an accountant for each of the increasingly agitated employees. This turned out to be a Godsend; for it gave all concerned a little clarity. KT, the accountant and I came up with a plan of action.

Since our silver bars would be hard to sell, we would turn some back to the company and get quick money to start paying our tax liability. Any coins that were too corroded to be easily sold would also go back for credit. We got IRS approved appraisers and had them re-appraise the items, hopefully to a more realistic value. We would market the rest ourselves, hoping to get more than what the company would give us if we had turned them back in. An amended W-2 would be submitted and tax would be paid on this more accurate, but still large sum.

For the next six months KT and I left Key West every couple of weeks for a long weekend of marketing. Dive shops in cities all over the south were quite anxious to have the exhibit at their stores and we borrowed items from Ted and others to round out our display. We picked cities where our friends or relatives lived so we could keep our costs down. To avoid unwanted attention, we drove from place to place in a nondescript car. The dive shops gave us the use of their glass cases, provided security and let us use their classrooms to present our continuously running slide shows.

We became quite successful at getting the media to cover the events; because they were free to the public we didn't have to spend money on ads. The day before our show we had a media day, doing radio, TV and print interviews. People would come by the thousands, and if they asked if any of the treasure was for sale (which they always did) we gave them the option. We only had two misfires; in one city a NASCAR race captured most of the media's attention, in another gale conditions dampened the turnout. Many of our "sponsors" asked us to come back a second time.

We sold a lot of treasure but mostly at fire sale prices. We were run-

A dive shop treasure exhibition. Photo: KT Budde-Jones

ning out of time and would soon be up against other competition that would be doing similar things. In the end there was enough money to pay our tax burden but the majority of our bonus was consumed in the process. The coins turned out to be a great asset for our niche marketing as most people could afford them. The large, uncut emerald was flashy but few had an interest in it. The stone would eventually be sold for less than the taxes we paid on it.

Dick Klaudt had to do his own marketing as well, so he and I coordinated to make sure that both of us wouldn't be gone at the same time. While KT and I had been gone, Dick had been forced to terminate an increasingly belligerent crewmember. With the ongoing heat and drudgery of the boatyard there were certainly fewer niceties between the workers.

Now it was my turn to be the heavy. The *Magruder* was big and complicated enough to require a full time engineer. Our current one had been with the boat since it was first brought to the Keys. Unfortunately, he REALLY didn't like the new management team. While we had been in the yard he had thoroughly stripped out much of the bilge pump manifolding, the pumps, the compressors, and had torn apart much of the other engine room systems to repair or replace them. We had let him pretty much do his own thing at first but he had become argumentative and secretive. The final straw was his abuse of a mild-mannered helper.

The engineer seemed to feel his position was secure since he was the

KT shows off some silver bars, hull spikes and silver coins she and the *Swordfish* crew found along the *Atocha* northwest scatter trail. Photo: Tom Ford

only one who knew the ship's systems. He was wrong. After years of studying and working under Ted's engineering expertise, there was nothing here that intimidated me.

Since the engineer had a number of firearms on board, I thought it might be prudent to have a cop on site when the termination occurred. The Monroe County Sheriff supplied a deputy, which helped defuse a potentially spicy situation. The good news is that the engineer went on to have a good life elsewhere, and our boat yard workforce appreciated the ease in tension.

As we got further into the boatyard project we found more and bigger problems. Rusted out bulkheads, hull plates, and leaking keel coolers kept turning up. The boatyard was *Arbutus* hot, and the additional heat from all the welding and cutting only added to the delight of the place. We were all seriously motivated to get back in the water and be done with this whole nasty project. Months of this still lay ahead.

KT was still working the *Swordfish,* and had been promoted to first mate when I left for the *Magruder*. She had obtained her captain's license years ago but this was the first official acknowledgement of her skills. Tom was a great mentor to her much as Ted had been to me, letting her do much of the boat handling and trip organization. This whole situation was much better for her, and since I was no longer in the loop, others couldn't assume that she was only there because she was my wife.

Tom, KT and the rest of the *Swordfish* divers worked aggressively further out along the northwest secondary scatter trail from the **Atocha Main Pile**. They had reached a zenith as a recovery team, locating twelve more silver bars scattered through an area the usually thorough *Dauntless* crew had just finished working. When Tom had to leave Key West, he told KT to captain the boat in his absence. Despite much politicking by the *Virgalona's* recent ex-captain, Mel, Deo and Ted put their full support behind her being in command. She and her crew did just fine.

On a later trip the *Swordfish* would make an unexpected discovery. While checking out a mag hit in Boca Grande Channel they found an intact World War II Avenger torpedo bomber in 35 feet of water. The aircraft's national markings were still visible, even though it was half buried in mud. Mel wanted it excavated, so they spent an interesting week digging carefully around it with airlifts. One calm day I played hooky from the boatyard and ran out to the spot for an inspection dive in our personal boat *Patience Pays*. KT and I had a long mutual interest in World War II history, and we were particularly excited by the find. It would later be determined that the plane had been lost in a training mission during the war.

With his dad's blessing, Kane Fisher had bought a barge and equipped it with living spaces and a large compressor. He anchored it over **Emerald City**, using airlifts to pump the ocean sediments to the surface and discharge them into a screened sluice system. It was manned by his own private crew who spent long hours hand picking out emeralds as they appeared.

Posing near the engine of the TBM Avenger   Photo: KT Budde-Jones

Looking for new and varied lost treasure projects absorbed much of Mel's time and imagination. With the *1622* ship's continued salvage in someone else's hands and money in his pocket, Mel could follow his lucky stars into the next big find. He spent a lot of time traveling. When in Key West he would advise Jamestown, although they seemed less interested in his opinions as time went on.

In our boat yard confinement, Dick and I always seemed to miss opportunities to engage the new owners. Others were taking a much more aggressive approach. Gaining early access to the new bosses was a golden chance to self promote, garner favorable considerations and settle old scores. The status quo equilibrium was about to be tipped.

We decided that we had better get into the game. With all the chest thumping going on down on the docks, we doubted that the new Big Guys were even aware of our existence. There was a difference now between us and the other boats. Mel hadn't sold the *Magruder* to Jamestown, he was leasing it to them. We were being paid by Jamestown but we weren't running one of their "company" boats. Our position felt more than a little precarious.

A friend from the office supplied the contact address and phone number so we could reach the principal partner. I called the number several times but there was never an answer. Dick and I wrote him an introductory letter, again no response. Things might not be going in a good direction.

In the midst of all the stormy company matters a real hurricane made a quick pass over Key West. Fortunately it was a very minor one that did little damage. The most memorable event during the whole thing was when Mel called during the few calm minutes as the "eye" passed overhead. He asked me to rent a plane and fly out to the site to check out the condition of Kane's barge that had been left anchored and unmanned on the *Atocha* site. When I told Mel that I wasn't interested in flying a Cessna forty miles out to sea in a hurricane he quickly replied, "It's not windy in the eye, just stay in it."

In time the new management began to assert itself. Kane was discharged from the *Dauntless* and Andy Matrocci was made captain. Duncan Mathewson was let go as well. Jamestown Treasure Salvors had hired their own ex-Navy man to oversee their interests in Key West. Unlike Ted, he had no background in marine engineering, salvage or diving, his last billet had been teaching sailing in Annapolis. Without any interview or discussion with the new owners, Ted would soon get the word that he was also no longer needed.

Ted was pissed off, and thought that the whole matter had been handled in a thoroughly dishonorable way. He believed that his termination was a sucker punch, instigated by a few of the divers who didn't like him. They had lobbied hard to shape the power structure of the new organization. It was a bitter payback for the effort that he had expended to get them a larger treasure bonus in the last division.

I had never known Ted to back down from a fight, but this time was different. Since the *Atocha* had been found, his interest and involvement in company affairs had visibly notched back several cogs. Marketing their own treasure and enjoying extended visits to Mexico had become more important now to the Miguels. Leah immediately quit her curator position, wanting to have nothing more to do with Jamestown. Mel and Deo were very upset by departure of their trusted friends, but could do nothing to change things. It was no longer their show.

By now the *Magruder* rebuild project had gone on for eight months. We were still finding problems, but Mel had finally run out of patience. He wanted us to button it up and get back in the water. Jamestown was getting nothing for its lease of the *Magruder* except more boat yard bills.

After a false start with a balky transmission, the *Magruder* finally made its first trip to the sites in mid-September. After nearly nine months of our sweat staining the marl of Stock Island, it was great to be back at sea and in operation again.

I had begun training a mechanically sympathetic welder/diver named Greg for the engineering position and had to attend to his questions as they came up. While putting all the ship's systems back together, I had gotten very familiar with them.

Our first trips went slowly as everyone was learning new routines. In the boatyard Vince Trotta had worked through his issues with me. As the most experienced crewmember, he had been promoted to first mate. We chased down some old mag hits and then explored some on the **165 degree line**. With KT on the *Swordfish* and me on the *Magruder*, we truly became ships passing in the night. While in port, we still spent our free time at the Miguels. Ted and Leah were our best friends but without their involvement in the company's affairs, gossiping about work was unfamiliarly one sided.

Artifacts were continuing to be recovered by all the boats but at a much slower rate than during the previous two seasons. There was also a subtle but noticeable split in the way the different boat crews interacted with the new owners. The *Dauntless* boys seemed to have the inside track on what was going on. A late season meeting with all the boat crews and the principals of Jamestown was announced. It would be the first time we would all be together at the same time with the Jamestown heavies. The *Magruder* would serve as the conference hall.

The big meeting was in two parts, the first with just boat captains and the Jamestown hierarchy, the second included everybody else. During their portion of the meeting, the captains inquired about the intentions of the new management. Were they making plans for research on other projects? Since the *1622* sites had already been heavily worked, we needed to start thinking about the next expedition while gradually winding down on the *Atocha* and *Margarita*. Our boats and their equipment were getting old; was there any thought about updates or replacements? The new bosses looked at us like we were speaking in tongues.

The second part of the meeting ended on a bizarre note. We were lectured on how many untold millions we would be finding soon on the *Atocha* and *Margarita*... despite what we had already found, their potential had barely been scratched. The company was now like a championship football team and would soon be the winner in the treasure Super bowl. We all watched quietly and a little embarrassed as our new Jamestown bosses chanted in unison like high school cheerleaders "We're number one! We're number one! We're number one!" No one felt compelled to join in, but at least one diver made the decision to quit then and there.

Jamestown's operations manager had found a new headquarters office for them near the Waterfront Market. They had now separated themselves completely from both the Museum and Mel's corporate headquarters in the old navy warehouse on Greene Street.

Things would get weirder as the end of the year approached. Something was up, but it was hard to interpret. The *Dauntless* guys seemed to know something but weren't talking. The new operations manager was cordial, but close lipped. On the *Magruder* we continued to work both sites as the weather allowed. Just before Christmas, Dick, KT and I all went off to visit relatives.

The New Year brought developments as subtle as a dinosaur-killing asteroid. Tom Ford quit, tired of all the years of hardship and hassles, he could see things only getting worse with Jamestown. As first mate and a licensed captain, KT would normally be considered his logical replacement, but Jamestown had other plans. They had formulated a brand new crew roster for their company boats. Greg Wareham was promoted to captain of the *Swordfish*. Vince had been recruited for the position of first mate from the *Magruder*.

The *Swordfish* and *Dauntless* would now tie up at their own docks instead of Mel's. Everybody else in the company not selected for the Jamestown boats was sacked. The *Magruder* and her crew were out of the loop.

There was more. Jamestown had apparently gotten frustrated when their predicted quick treasure bonanza on the *Atocha* and *Margarita* hadn't materialized. Their response was to drive off a cliff, marshaled with the help of a local dirtbag.

Sam Kirby was an aged expatriate Englishman who had long ago taken up residence in the Key West area. He existed on the edges of homelessness, not quite living in the mangroves, but not exactly thriving either. Sam looked and sounded like a character out of a Dickens' novel, bald except for a fringe of greasy gray hair, a large hooked nose and a complexion that only comes from too much outdoor exposure. His clothes were unwashed and there was a definite smell about him. He was often seen wheeling grimly along on his beat up adult tricycle.

I had met old Sam years earlier in a Stock Island restaurant, and had quickly concluded that he might not be wound too tight. He was widely known by locals between Stock Island and Key West as yet another eccentric but basically harmless "special" person. Key West's easy going personality often enabled rather than expelled oddballs. Sam continuously spewed out that Mel was a fraud and that only he knew where the REAL *Atocha* and its treasure were.

Jamestown committed suicide in the biggest way possible. First, it abandoned the *Atocha* and *Margarita* sites completely, stopping payments to Mel. Secondly, through kismet they had hooked up with old Sam Kirby and had fallen in love with his tales. Sam's usual stories of guardian Jesuit priests and how he had personally swum down into the "real" (and completely intact) *Atocha*, opening the cargo hold door to view the stacked treasure stored inside were apparently irresistible. For months Jamestown's boats followed Sam's psychosis further into the financial toilet.

Despite welching on their purchase agreement with Mel, Jamestown's purse strings remained open. Both the *Dauntless* and *Swordfish* crews quickly acquired lots of extravagant toys in the form of big stereos, underwater scooters, semi-automatic weapons and new, high-powered Whalers. None of this mattered in the end. When nothing came of the Sam silliness, Jamestown abruptly folded its tent. The boats were hauled out of the water at the Peninsular boat yard on Stock Island for indefinite storage. All employees were terminated.

Mel was as surprised as anyone at Jamestown's actions. He knew that the *1622* site's abandonment by his buyers might present an opportunity for outsiders to garner their ownership in Admiralty Court. Though he had enjoyed spending his time searching for new treasure hunt possibilities, Mel realized that the *Atocha* and *Margarita* were still his fame and fortune powerbase. Their association gave him credibility on any new venture with both the media and potential investors.

After three weeks of trying to figure out Jamestown's intentions, Mel decided to fund the *Magruder's* return to the sites. Dick and I rehired those of our crew who hadn't wandered off including KT, who graciously agreed to replace the departing Greg as engineer. We were the sole presence on both *1622* wreck sites.

Numero Only

Magruder working into the evening hours

The *Magruder* was now alone in carrying Mel's flag on the *Atocha* and *Margarita*. On paper we were now working for Cobb Coin Inc., Mel's personal company that ran his *1715 Fleet* salvage operations. After years of non-stop competition between the various company boats on the *1622* sites, it was quite strange to now be the entire operation. We would have to do everything, from the exploratory surveying of new areas to the continued recovery in the known ones. Dick and I were anxious to search for still undiscovered parts of the wreckage trails rather than perpetually hammering away at the known ones.

One of Mel's own theories would have to be satisfied before we pushed ahead. His immediate project required our moving many tons of ballast several hundred feet from our **Main Pile** ballast dumps. He was afraid that artifacts had been accidentally buried when we had first moved the stones. None of us were particularly thrilled about this idea, but were grateful enough to have our jobs back that we completed the task quickly and thoroughly. Nothing new turned up.

Dick and I soon shifted the *Magruder* up to the northwest trail that lead from the **Main Pile** in the direction of **Memorial Day**. Here we found lots of intact barrel hoops, potshards and silver coins. There was so much material on the bottom that we had to use lift baskets to recover the fragile yet bulky items.

It was my turn to dive; everyone else on board had already run out of bottom time. I was just getting over an unpleasant spell of the flu that had kept me out of the water for two weeks. A brisk wind imbedded in a mid-February cold front was still in the area. I chose to wear a dry suit over my clothes and full-face mask so I wouldn't have to get chilled or wet. As lunchtime was quickly passing, I wolfed down two hot dogs while putting on my gear.

The dive was uneventful; I was putting artifacts that had already been located, tagged and mapped into the recovery basket. I did a final sweep with a detector around the area to make sure nothing had been missed and headed slowly for the surface. The dive had ended an easy 15 minutes before reaching the no-decompression limit for that depth.

I waddled up the dive ladder and gave the signal to the crew on the back deck to start winching up the full artifact baskets from the bottom. As I took off my gear, a radio call came through on the back deck loudspeaker. Since the call required my attention, I trotted up to the bridge to answer it. At the end of the conversation, I continued stripping off my half removed dry suit.

During the undressing contortions, I developed a strong stomach cramp. I cursed myself for my bad eating habits; those slammed, pre-dive hotdogs must be restless. Very quickly the pain threshold passed anything in my experience, and at the same instant my legs went numb. I fell backwards onto the wheelhouse floor, no longer able to stand.

By chance First Mate Tony Absetz and KT both arrived in the main salon at the same time through different doorways. She had just come up from the engine room and Tony stepped in from the back deck. Through the passageway that lead up to the bridge they both saw me go down and ran to see what the problem was.

By the time they reached me I could feel nothing of my legs and felt like my waist was being cut in half by a chain saw. An attack of decompression sickness (the "bends") seemed to explain what was happening. All three of us shouted "Oxygen!" simultaneously.

The *Magruder* was equipped with an oxygen bottle for just such decompression accidents. Within five minutes of its administration the pain had largely subsided and I could feel my legs again. By this time Dick and the other crew were aware of the situation and made a radio call to the Key West office. I could walk again and lay down on the couch in the salon. I was feeling less pain by the minute but was very tired.

It was understood around Key West that the Coast Guard was usually overtaxed and was often slow to respond. The office called Ted, the only person they could think of who had the resources to deal with the problem immediately. He grabbed another oxygen system, got his personal boat underway and rushed out to the site. The hospital was notified and arrangements were made for the ambulance to meet us at the dock when we arrived in Key West.

Dick and the crew quickly recovered the anchors and headed to Key West to lessen the distance Ted would have to travel. They continued to port after I was picked up. We met Ted in Boca Grand Channel where I transferred into his boat. By that time I felt normal and a little silly, but I used the oxygen for the rest of the trip.

The ambulance was late and I sat at the dock for a half hour before it arrived. The vehicle's own oxygen system was empty, and the driver and attendant got into a shouting match over whose responsibility it was to check for such things. I broke up the argument by reminding them that the hospital was waiting for me.

At the hospital I underwent x-rays and observation for several hours, all while I still breathed pure oxygen. Deo, Ted and Leah stopped in but were not allowed access. After being visited by several doctors I was finally released. I was surprised that I wasn't being sent to a hyperbaric chamber for further treatment, but was confidently told by the doctors that they could find no reason for it. I asked the doctor when I could dive again and was told, "Tomorrow, I guess." This sounded too good to be true. By this time the *Magruder* had reached port and KT came to take me home. I was exhausted and fell into bed.

The coconut telegraph had been very busy; word had spread quickly through the island's diving community that I had gotten "bent." Tom Ford called, wanting to know if the hospital had put me in a decompression chamber. When KT told him no, he became concerned and related the story of a friend that had been given the same inadequate treatment and had suffered serious problems because of it. Ironically, as Tom called I was just starting to feel pain again.

Tom and KT made arrangements through the Diver's Alert Network for me to go into a decompression chamber in Miami. Foregoing another ambulance experience, KT drove me there in our own car while I breathed from an oxygen cylinder. Six long hours in the chamber seemed to cure my problem.

The attending hyperbaric doctors could not explain why I had gotten the hit. "Sometimes things like this happen, perhaps the flu changed your metabolism. We may never know the exact reason. Yours is not the only case like this we've seen."

After the treatment three separate hyperbaric doctors who all had different opinions about my vocational future interviewed me. When I asked them when I could begin diving again the first one said "Never." The second wasn't sure if I should or shouldn't. The last one was more optimistic. "I don't see any reason why you can't, just stay out of the water for at least a month and be VERY conservative on the dive tables." I liked his answer best.

Mel soon answered the question of what I would be doing next. His latest Caribbean misadventure was a partnership with Brazilian Tony Kopp.

Tony was a skilled conservator and wreck hunter. Mel had once hired him to help develop the wreck site of the slave ship *Henrietta Marie* that had been found during the early *Atocha* surveys. When the *Atocha's* **Main Pile** was found, Tony had worked in the conservation lab helping to stabilize and preserve the artifacts. Now they had combined forces on a new venture in the waters off Antigua and Barbuda.

This new project sounded like a reprise of the Vieques debacle. He was betting that numerous interesting and rich wrecks would be found, and investors would find them irresistible. A museum would be built on Antigua to satisfy the local government and showcase our many finds. Cheap property available on the underdeveloped nearby island of Barbuda would be a great place for Mel to build a huge, world-class hotel resort in which his investors could stay. People would come from around the planet to stay and dive in this ultimate treasure hunter's paradise.

Mel contracted a large aluminum crew boat for the new expedition and Tony thoroughly outfitted it. The vessel was delivered to Antigua, ready to find and recover artifacts from colonial period wrecks. Unfortunately, this project suffered from some of the same conceptual defects that Vieques had. There were no specific targets of interest, just the supposition that it "Looked like a likely place" for old shipwrecks to have occurred. Equally serious was the lack of investor support for Mel's latest "wildcatting" venture.

Since Tony had other obligations and couldn't stay, Mel thought of the perfect solution; he would send you-know-who to Antigua.

The *Shoreline 21* anchored in St. John's harbor

We had already been warned by the office bookkeepers about how little investment money had been raised for this new expedition. That was scary enough, and with our knowledge about the lack of notable recorded wrecks in the area it was hard to muster any good feelings about our involvement.

Once again we tried to beg off, but Mel was inflexible. Dick would have to run things on the *1622* sites until we came back at some unknown time in the future. KT and I reluctantly departed again, this time with open return tickets just in case the promised funding failed to materialize.

Our base for this new project was St. Johns in Antigua. Our expedition boat was the *Shoreline 21*, a giant, nicely equipped version of the *Swordfish* that would have served perfectly on the *Atocha* site. Sadly, due to lack of fuel money, it spent most of its time anchored in the open sewer harbor. A black Englishman named Kevin currently lived on board. He made an amusing and helpful companion for the few weeks he was on island. Tony Kopp met us at the airport, briefly checked us out on the boat and showed us locations of several sites he had confirmed. Soon enough he had to leave and we were on our own.

Spending time anchored in St. John's harbor was pretty disgusting. The harbor was an unflushed toilet garnished with a neverending dump fire of old tires burning along the rim. The island probably looked a lot more appealing from the vantage point of a private, $500-a-night hotel compound, but from our perspective it was just nasty, nasty, nasty.

Fortunately, the boat had a generator to operate the air conditioner. We had to keep the doors tightly closed to keep out all the flies. More of a concern to us was our fresh water source. The only place in the harbor to get water was at the fuel dock, and then only if you bought fuel. The boat had a desalination unit on board that processed seawater into potable water for showers, drinking etc. Ever mindful of the kind of water in which we sat; we REALLY had to believe in the technology.

I was a little nervous about diving again. It had been over a month since the hyperbaric treatment and unfortunately my first dive since the accident would be in the Third World. We checked and found that there were no hyperbaric chambers on island. If I suffered another incident here, well… it would be bad. At first I was very conservative, diving only to shallow depths and returning to the surface very slowly. Since nothing evil happened, I gradually worked deeper and longer.

As we worked the wrecks off shore of Antigua, we found a number of archaeologically interesting sites. We particularly enjoyed working the scatter trail of a 1700's English merchantman in Deep Bay. After a predictably rough passage to Barbuda, an incomplete magnetometer survey among the reef systems produced interesting diving, a few anchors and little else. The mag with which we had been supplied was little more than a toy and the boat's low fuel reserves limited the amount of time we could spend there.

The real drama of our Barbuda stay was a rapid Felini-like juxtaposition of events: A crazy Englishwoman living in a roofless hotel ruin, sand flea bitten Italian tourists in their high fashion dress clothes crying on the beach, raisins flying off raisin bread in an open air bakery and half naked islanders who wanted to rent us their Land Rover. Some of the locals decided to board our boat while it was tied up to the only dock on the island. With machetes they tried to hack into the living spaces and shouted how they were going to kill us. The boat's metal construction with internally secured doors put a damper on their fun.

In the end, the project tanked from lack of money; the investors Mel had hoped would come through didn't. KT and I tried to keep the boat operation going, but our operational radius shrank down to just one site right outside of St John's harbor. There were only just the two of us left on board and with no money for fuel or groceries our options were running out. When the fuel tanks got down to the last few gallons of diesel we were in trouble. No fuel equaled no generator and without it there was no fresh water or electricity. When our calls to the office confirmed that no money was available, we anchored the boat in the middle of the harbor, shut down the systems, locked it and headed for the airport and Key West.

Mel still hoped to keep alive the Antigua/Barbuda project. He eventually sent one of the *Magruder* divers down to babysit the *Shoreline 21* while it sat in the harbor; a long distance *Arbutus* duty. When the *Shoreline21's* commitment ran out, subcontractors tried to keep things going but the operation died from lack of money. Mel continued his wildcat quest elsewhere...

KT and I got quickly back to our normal duties aboard the *Magruder*. We had been in the West Indies about six weeks. During that time Tom Ford had married Greta Phillips, a talented and gregarious woman who multitasked as an accountant, treasure salesperson and gift shop manager at the office. We were sorry to have missed the event.

The *Magruder* world hadn't changed while we were gone, it was still blue, thrumming and gymballed. Painted in the colors of sky and sea, its hull and even the interior spaces were decorated with tints of blue. The *Swordfish's* machinery noise always had a brutal edge to it, but the *Magruder's* was a softer but perpetual *Basso Profundo* of muffled generators, engines and other equipment. The massive fuel and water tanks built into the sides of the hull accentuated the boats tendency to roll, even in calm seas. New crewmembers found it hard to walk without grabbing onto something solid. The AC system efficiently circulated cold, diesel fuel perfumed air through every compartment - a real thermal shock when crewmembers stepped inside from the summertime back deck blast furnace.

Artifact conservation projects in the lab   Photo: KT Budde-Jones

Dick and I continually worked on both *1622* sites as best we could but the weather was unusually bad. There wasn't a great deal of money in the coffers for our operation and we struggled to make each trip productive. We found more artifacts despite the uncharacteristically rough summer seas.

Crippling us further, our Del Norte system began failing regularly from age and use. I tried to keep the system going by swapping out what extra parts we had left, but this was only a stopgap measure. Without Fay Feild's diligent maintenance we were forced to use the old, inaccurate LORAN as our navigational reference. This was a completely unsatisfactory situation, putting us back into the dark ages. Fortunately, Mel soon hired Jacques DeSchmitt as our in-house electronics repairman. Though not as skilled as Fay, he kept some of the old equipment functioning, at least for a while.

Our crew had stabilized and matured, they were experienced and careful in their work. Dick, KT and I imparted what wisdom we had to the newer guys who really seemed to listen. Tony Absetz functioned efficiently as first mate. This is not to say that there weren't personality problems; some of the others still didn't like a woman being on board, especially since KT's engineer status gave her authority over them. Except for the unusually crappy weather, shipboard life was as good as it ever got.

We turned our artifacts from each trip into the archaeologist-conservator who currently worked in the lab. The lab looked nothing like the beehive of activity it had been only a year ago. During the Jamestown Treasure Salvors slump, all of the lab employees were terminated. Mel retained one of Duncan Mathewson's **Atocha Main Pile** hirees to keep things going but despite his towering credentials, he was very disorganized and absorbed with personal projects. One electrolysis tank was operational and the lab was strewn with garbage and broken equipment. Artifacts in wet storage awaiting cleaning seemed to be piling up everywhere. One could barely walk through the lab because of the containers and trash. We *Magruder* boys started to have concerns that the artifacts the company was paying us to find were not being dealt with in a timely fashion.

The problem was solved when the archaeologist moved back to Canada. Dick and I decided to use our port time to straighten up the lab and inventory all of the artifacts waiting to be cleaned. We brought in the whole *Magruder* crew to participate and over the next weeks we cataloged thousands of items that had yet to be cleaned. Most of the objects had been awarded to the company in past divisions and had drawn back burner status so that the investor's pieces could be stabilized and cleaned first. We made sure that the recently recovered artifacts were all in process. Whenever we returned to port, our crew split up to work both on the boat as well as in the lab.

Office bookkeeper Sandy Dunn and her husband Phil decided to sell their new townhouse for a life exploring the Caribbean on a sailboat. Sandy had been a good friend and confidant over the years and we would miss her. Not long after hearing the news, KT and I stopped by their place "Just to check it out." We were impressed with the house and were surprised to find that they were selling it complete with all new furniture, TV's, linens, etc. Though we had not thought about it before, the idea of buying the place seemed to have merit. We brought in Ted to give an assessment of the situation. In his own economical and decisive way he simply said, "If you don't buy it, I will." We shortly became Key West property owners.

Mel's next wildcatting adventure landed hard in his lap. A toothless older man named Grayden and his dog-collar wearing daughter showed up late one day at Mel's office. The man stated that he had found the fabled gold ship *Central America* off the coast of North Carolina and that he had a video and the research to prove it. Mel viewed the video with Grayden's commentary, and with only a casual glance at the research decided that this undoubtedly was to be his next big thing. Grayden had general coordinates for the site. He admitted that he couldn't dive and had no boat, but he would be willing to share in the tons of gold if Mel wanted to kick in his equipment, personnel and expertise. Mel excitedly wrote up a contract that they both signed.

Mel summoned me to his office. With the intrigue of a secret military briefing, as I closed the door, Mel gave me a sealed envelope containing my operational orders. Mel told me that the envelope was not to be opened until we cleared the sea buoy south of Key West. All he told me was that we needed to fill the *Magruder's* fuel tanks and provision her for at least a month. He told me that I was to take along all the equipment we needed for an extended search and recovery. A non-company employee would be joining us on board for the trip and would act as a guide when we got to our destination. Mel did not want to give out any more information than that; he intimated that the nature of the project was extremely sensitive. If anyone found out how easy and accessible it was we might be beaten to the punch.

I was irritated about having to once again stop working the *Atocha* and the *Margarita* sites; we had just started to cover some ground on both wrecks. Although the backcountry looks and nature of our "guide" raised doubts, we were in it for the duration.

When I asked about the legal status of this "project", Mel was very insistent that by the time we arrived on site, the Admiralty claim paperwork would have been forwarded on to me. Since it would take time for the slow moving *Magruder* to get where it was going, I only had 24 hours to get underway. Though the *Magruder* was the vanguard, the long idle *Bookmaker* would be joining us in several weeks as well. Jacques LeMare and his girlfriend Terri, a vagabond couple that Mel had recently befriended, would bring it to the location.

The next day, after hurried preparations, we steamed out the south channel towards open sea with our mystery guest Grayden on board. The rest of the regular crew was excited about a new adventure and full of speculation about our destination. All we knew was that it wasn't a foreign adventure. I had been told that we didn't need passports.

For this expedition I was the sole captain, Dick decided to defer to my seniority. Since this was unlikely to be a short trip, we had devised a schedule that put Dick, Tony and KT each in charge of their own four-hour watches. Each watch officer was to oversee their own lookouts, helm, navigation and engineering. I would not stand watch, but would be called to the bridge at any time during bad weather or any unusual problems. Our guest was just along for the ride until we approached our destination.

As we hit the sea buoy I opened our orders and read them to the crew. We were to steam to Beaufort, North Carolina where we would find the gold ship, the *Central America*, which Grayden had discovered. We would establish operations and locate the wreck under Grayden's guidance. We were to recover an artifact from the wreck so that our lawyers could file an Admiralty Claim for the site. The *Bookmaker* would be joining us to assist with the bulk recovery of artifacts.

I plotted the course for our destination and gave headings to the standing watch. Grayden was anxious for me to look at his research and video, so as soon as my duties allowed, I took the time to watch the video, read the documentation and questioned him further.

The video was an unedited, amateurish production that showed a rusty triple expansion steam engine sitting on a sandy bottom in fairly shallow water. Grayden had accumulated stacks of research on the *Central America*, most of which was photo copied from library books. Although I had a vague knowledge of the ship's history, I had never studied it. The research provided much documentation on the construction of the ship, its engines, passengers and cargo. At the time of its sinking, coordinates taken from other vessels that had rendered assistance to the *Central America's* passengers. It took me over an hour to digest all the information.

I immediately found problems. The coordinates taken at the time of the sinking were hundreds of miles from were Grayden had indicated he had found his wreck. The water depth at the original recorded position would have been thousands of feet deep; Grayden stated his site was in about 35 feet of water. Granted, the original coordinates were dead reckoned in hurricane conditions, but they certainly shouldn't have been that far off. The steam engine on Grayden's video was far too small and of a completely different design from that of the *Central America's* propulsion system. He hadn't in any way tried to correlate his research with the physical evidence.

When I pointed out to him that his own documentation didn't support his claim that this wreck was the *Central America*, he was stupefied. His only reply was that his grandfather had once told him that a body, thought to be from the ship, had washed up in the Beaufort area. I told him that North Carolina was called the "Graveyard of the Atlantic" for good reason; there were many hundreds of ships sunk along the coast, especially in the shallow areas. The strong Gulf Stream current could have moved a floating body some distance from where a ship sank.

After I interviewed Grayden, I began to get the real picture. Fishermen had found an old steam engine in shallow water, and through chance Grayden found out about it. He immediately decided that it must be the *Central America* and paid divers to make a video. Hearing about our finding the **Atocha Main Pile**, Grayden made the long trip to Key West in the hope of enlisting Mel's help. Mel, desperate for a new wildcat adventure was already primed when Grayden made his case. The last act of this *Opera Buffo* was the *Magruder's* present trip. We would live out the consequences.

The word quickly spread throughout the crew about the now tarnished legitimacy of Grayden's story. Despite the farce, we all knew that Mel would never be satisfied until we had thoroughly disproved the issue by physical inspection of the site. Our long boat ride just got longer.

As bad as the news was about our mission, I felt a surprising exhilaration. As the *Magruder* made its way though the deepening twilight, I stood alone on the bow. The vibrating steel deck beneath my feet was part of a good ship; the people on the bridge, in the galley or in their bunks were a fine crew. I had been given the resources and orders to find shipwrecks on an ever-expanding scale. Here we were again, steaming off to unknown adventures embraced by the ever-changing dynamics of the sea. Soon we would be out of radio range of any land stations and the handling of all events; good and bad would be a test of my experience and authority. I was ready.

There is an intercoastal waterway that allows shoal draft vessels protected passage along much of the U.S. east coast. Since the *Magruder* drew more water than the depth of the waterway, we had to make our trip entirely off shore. To boost our twelve-knot speed we went far out into the Gulf Stream where the current added another three knots or so to our pace. Ships of all sizes and speeds routinely used the Gulf Stream as an express lane while traveling north, south bound ships came in closer to shore to avoid the current. At night the many navigational lights that moved along these two shipping lanes often resembled a freeway.

We ducked into Charleston for a couple of days to get out of some bad weather. During the trip we had noticed that Grayden's complete wardrobe was the same T-shirt and cheap red athletic shorts with which he had boarded the boat. He eventually did turn his shorts around after a few days and then inside out so all sides got dirty. Watching him eat was not for the squeamish either; somehow he managed to quickly consume steaks and corn on the cob despite not having any teeth. One could always tell what he had eaten as his cheek whiskers and ear hair was flecked with his most recent meal.

After the storm blew out we made it all the way to Beaufort, found a dock and a nearby phone to report our arrival to the Office. The *Magruder* and her crew had performed flawlessly; everyone had adapted to our non-stop schedule and done a great job at their assigned watch duties. Even Ted would have been impressed.

We quickly got down to business and went out to find Grayden's folly. The coordinates he provided had gotten us into the right area. Our fathometer soon showed a sharp spike projecting from the bottom. KT and Tony dove on the anomaly, finding it quickly despite the horrific current. Half immersed in sand was the same steam engine as seen on Grayden's video. I anchored the *Magruder* directly over it and we all took a look.

Sure enough, this wasn't the *Central America*. It not only was the wrong type of steam engine, but the dimensions we measured weren't anywhere near close to the size of the *Central America's* propulsion system. A detector survey showed the wreck to be quite small. As far as I was concerned we had wasted enough time and money on this lame exercise.

Grayden's enthusiasm wasn't diminished; he stubbornly wanted us to start bringing up the gold. All I wanted to do was to go back to the dock and call Mel to report that this wreck was not what he had hoped. The crew was impatient as well, ready to chuck Grayden and the whole adventure. We all wanted to head back to Florida to salvage what was left of the summer season.

Mel told me over the phone that he wanted us to excavate around the wreck just to check it out. If nothing of note was found, he reluctantly agreed that we should abandon the site. He mentioned that although Admiralty claim paperwork was already in the pipe, there had been an unexpected delay as our normal attorney, David Paul Horan, was not able to take on the project.

Pat Clyne, who arrived by car, had come to film Mel's latest triumph. Pat's position with the company over the years had evolved into being the Fisher family's personal videographer and P.R. person. He was fiercely loyal and dedicated to Mel and family; he always presented them and their activities to the world in the best possible light. Some of the more cynical members of the crew saw his job as simply "perpetuating myths". Mel wanted to be known as "The World's Greatest Treasure Hunter" and Pat was there to make it happen.

I explained to Pat that the wreck wasn't the *Central America*, there was no gold, and Grayden needed to go away. Pat hung around for a bit until we made our next move. He told us that the *Bookmaker* was already enroute to our location. Mel had told them to bring shovels so they could help us more efficiently in loading the gold.

When the changeable weather allowed, we revisited the site and excavated it as Mel had instructed. As we had already discovered, it was a small wreck that appeared as if someone had previously partially salvaged it. There were few artifacts and we did not disturb them.

By now Grayden had really worn out his welcome. In addition to wasting our time and money, he was quite ripe from not bathing the whole trip. The final straw came after we had proved conclusively that this wasn't the *Central America*. Grayden called Mel and told him that if this wasn't the *Central America*, it must be a Confederate blockade-runner loaded with gold.

KT became the point man of our collective anger by cornering Grayden in the phone booth. She made him understand in no uncertain words that he should immediately disappear into the back woods where he had come from and never again contact Mel with his dopey crap.

In reporting our final assessment to Mel, I finally convinced him that there was no further point in our lingering in Beaufort. He told me to head to Ft. Pierce and the *1715 Fleet* sites. There were deeper areas there he wanted us to explore where the small local boats had been ineffective. The *Bookmaker* was recalled as well and would join us there.

Everyone wanted to get back to business on the *1622* wrecks, but a few weeks on the *1715 fleet* might be an interesting interlude. It certainly beat what we were currently doing. We quickly booted Grayden to the dock, hoping that he would fade away rather than stirring up Mel with more of his folksy B.S.. Despite a weather forecast of an approaching powerful storm front, I decided to get underway immediately before Mel could change his mind. Once offshore, we would be out of easy radio contact. We left Beaufort late that afternoon.

By ten o'clock that night a full-blown gale overran us. The wind blew from the north directly against the Gulf Stream current, and the waves were stacked up tight and high. By midnight, things got worse. Every time I made my rounds into the wheelhouse there were anxious faces looking at me for assurance.

We progressively extended our double towing lines to accommodate the *Whaler* we were pulling behind us. Caught by the huge waves rolling up our stern, the small boat was often shoved violently right into our transom. After we let out over two hundred feet of towline, the *Whaler* began to surf next to us almost even with our wheelhouse before the *Magruder* itself was carried forward down the hissing wave face. The towlines sometimes snapped hard as the *Magruder* rocketed away from the smaller boat. This cycle was repeated endlessly over and over again.

The wave heights were impressive. Even though the *Magruder's* bridge windows were nearly thirty feet above the waterline, in the wave troughs all we saw straight ahead was the rumpled back of the preceding wave. About one in the morning a passing freighter about a mile off saw our lights and hailed us on the radio. After confirming positions and ship names the radio operator queried, "*Magruder*, what's your length?" When I told him he replied; "What are you doing out here? We're four hundred feet long and we're getting our asses kicked!"

Nothing good ever seems to happen at three in the morning. The standing watch reported that the *Whaler's* heavy-duty main towing eyelet had snapped off; they had retrieved the broken piece still attached to the now useless towing rope. The secondary line was attached to a much less substantial cleat on the top of the bow. It wouldn't last long in these conditions.

We spent many hundreds of manhours modifying this *Whaler* for our purposes; it couldn't be easily replaced. It was so integral to our operation that we would be completely out of business if the boat were to be lost. Since the company's purse strings were already pulled hyper-tight, there was no ready replacement money.

The fragile secondary tow cleat only gave us minutes of control over the *Whaler's* destiny. When that final umbilical parted, the black raging seascape would quickly swallow up the craft.

The waves were so big that the Magruder's slo-mo responses put it at risk of rolling over if I tried to turn it back into the seas. Things were getting away from me; maybe they already had. Early in the trip I had felt ready for a challenge. Now I had it.

After a quick evaluation of the wave cycles, I briefed First Mate Tony Absetz on my plan. "This is going to be really dangerous. If the *Whaler* gets away from us we're out of business - I think you know that. At the right moment I'll need to bring the *Magruder* around 180 degrees into the wind so I can hold position. That's going to be bad enough. There's no way that we can bring the *Whaler* up astern of us so you can get in it, because the mailboxes would crush you. We'll have to bring it alongside. Timing will be everything; you'll have to pick the right moment to get aboard. There are two good inward-facing cleats about a third of the way back from the bow, if they pull out they'll take the whole bow with them."

"I need you to make a bridle to attach to them so they can assume the main strain of the tow. Once that's done you'll have to pick your moment to get back aboard. I'm going to be pretty limited in my ability to maneuver and I can't let us get sideways while this is going on, we'll roll over. If you were to fall over the side it would be a challenge to get you back. Are you up for this?"

I picked Tony for this adventure because he was quick, athletic, had good sea sense and carried out instructions well. He also understood the gravity of the situation. Tony indicated he was ready to go. While he and the others got the necessary equipment ready I got ready to turn the *Magruder* back into the seas.

After watching the wave cycles to pick my moment, the time came. I gave the announcement over the P.A. for all hands to be on deck - we were making our turn. Just as the crest of one wave peaked under us I slammed the throttles wide open with one engine in forward and the other in reverse with the wheel hard over. Even as the ship vibrated hard from the strain it turned oh-so-slowly. We had just about made our full turn when the tall peak of the next wave pushed our bow toward vertical and then crashed over it.

So far, so good. I held the *Magruder's* position into the seas. The crew pulled the *Whaler* alongside and Tony waited, ropes in hand for his moment. He chose well and made it into the tossing boat. I pulled the *Magruder* forward away from it so the larger vessel's movements wouldn't crush him. While he worked at the new rigging, the secondary towline remained our only connection. We shone bright lights into the spray-fizzed darkness so Tony could see. His lifejacket had a small battery-powered light, which looked so small and dim in the maelstrom. Would we be able to find this light if he were washed overboard?

Tony quickly completed his task and was ready to be brought aboard. With the *Whaler* once again alongside he made the precarious leap up the lowered dive ladder away from the danger. After we were rigged again for towing I calculated the best moment and took the *Magruder* through another gut-tightening one eighty degree turn to run again with the seas. The operation had gone perfectly, everyone, especially Tony, had earned their corn flakes. It had been close; if the waves had been a just little closer or higher the *Magruder's* turn into and away from the wind would have been a disaster. We had been lucky.

With dawn it became apparent just how big the waves had become. We put into Charleston and made permanent repairs to the *Whaler*. After several days the seas calmed enough that we again began our trek south.

Just Beyond the Horizon

It had been over a month since our last trip to the *Atocha* site. In September the seas usually picked up with the first cold fronts of the year, preventing access to the shallow, shore hugging *1715* wrecks. Our arrival in Ft. Pierce was timely; the weather and sea conditions were still good, which gave the *Magruder* crew time to visit some of the local sites.

The good weather held unusually late, which allowed us to work into the last of October. Thanks to some good tips by John Brandon, we found artifacts on all three sites we visited. Our real surprise was on the *Cabin Wreck*, the first of the *1715 Fleet* to have been found in modern times. We worked just off the beach from Kip Wagner's original cottage where, in the early 1960's, Kip had first noticed old coins washed up on the shore. Since then many dozens, if not hundreds of modern salvors had recovered artifacts from this spot. In slightly deeper water than was usually worked, we found hundreds of coins and other artifacts. It was a nice bonus for us after all the hassles and trauma of the Beaufort bad juju.

Most of the wreck scatter was in water less than 15 feet deep; with our mailboxes in the down position we drew 15 feet. To use them we had to work a little seaward and the divers had to take constant care to make sure that as the boat moved up and down in the swell they weren't "cookie cuttered" into the sea floor.

Because of sharks, the beaches were closed to swimmers. This time of year huge shoals of baitfish swam in close to the shore, drawing schools of hungry sharks in their wake. With water visibility near zero, sharks would occasionally mistake an arm or leg for a baitfish snack. When we worked in

Silver pieces-of-eight and a gold coin fresh off the bottom found by the *Magruder* on the "*Cabin Wreck*" Photo: KT Budde-Jones

the shallow, turbid water, terrified baitfish escaping the jaws of hungry predators often pelted us. One hoped that bump just felt was the other unseen diver working with you in the dark hole.

We later learned why Mel's normal attorney, David Paul Horan, hadn't been involved in the Admiralty paperwork for Grayden's *Central America* fantasy. David had already been retained by a group that had actually found the real wreck. He couldn't tell Mel about his involvement or knowledge of the find until all legal issues had been resolved. As Grayden's library research had implied, the actual shipwreck was found in thousands of feet of water, near the original coordinates of the sinking.

By mid-November 1989 we were back on the *1622* sites. Over the months Dick and I guided the *Magruder's* path of search and excavation between the *Atocha* and the *Margarita* as weather and finances permitted. They were pleasant trips, with enough emeralds, coins and bits of gold jewelry found to balance out the usually dry and tedious sojourns through unknown sectors. Ashore, events were definitely more toward the uncharted.

KT and I visited the Miami boat show. We still shared the fantasy of buying our own live aboard sailboat in which to tour the Caribbean. After our window shopping pick was made we ran into a sailing school named Women for Sail that promised to teach women how to master sailing skills on similar sized boats. Jill London, the owner, had started the school after experiencing

Interesting finds still continued for the *Magruder*, such as these uncut emeralds, gold rosary beads and the gold medallion found near the *Atocha* "**main pile**".

the trauma of her husband's attempts to train her. KT signed up for the school to get a feel for the sailing world. Upon attending the class she found that she had a bigger captain's license than her instructors did, and that all boats are the same from the gunwale down. She still had a great time.

In port one day Dick surprised us all. He and his wife Heather had decided to leave Key West for an opportunity to build and run a restaurant in the northeast. They had always appeared so "established" in Key West. Heather seemed to know everyone on the island and was a longtime employee at the Key West Halfshell Rawbar. She and Dick were almost unique among the T.S. operations staff in that they owned their own house. They sold their place and moved on.

Ted had been complaining for a while of prolonged stomach problems. At first both Leah and Ted were mystified, as Leah said, "Ted's the kind of guy who gives ulcers, not gets them." The doctors didn't seem to have a clear idea either, and Ted's frustration grew as many visits and tests led nowhere.

Eventually the doctors made a diagnosis. Ted had pancreatic cancer. KT and I didn't know enough about the condition to be really scared. Leah told us the doctors thought that a cure was possible, that the cancer was very localized and that surgery would be a good option. None of us who knew Ted thought that a tiny isolated cancer would be enough to knock him down. We hung on to that thought.

Ted went through his surgery and despite post-procedure discomfort, life improved gradually for him. His appetite came back a bit and the pain subsided but did not go away completely. The combination of his condition and the operation made him much more fragile and older looking. For the first time, he lost his aura of vitality. This more than anything concerned KT and me. Leah kept her brave face and we all nervously hoped.

While we lived in Australia, the intelligence and beauty of the Sulpher Crested Cockatoos that we saw there had intrigued us. An opportunity came for us to get one locally and with more emotion than logic, we "adopted" a one-year old baby. When we went to sea, the poor feathered guy (named Dinkum), was immediately indoctrinated into life on the boat. He coped with us and his new situation pretty well, probably better than the crew did over his noisy ways.

Our nephew, Rick Lueneburg, came to visit us in Key West and decided to stay. For a while he served as a diver on the *Magruder* and fit in well. In his usual paternalistic way, Ted helped expand Rick's universe even further by giving advice on other job possibilities. Often Ted and Leah came over to our place for dinner. It was an uneven payback for the many hundreds of meals and good times that we had shared with them at 417 Cactus Drive. For a while things were almost normal.

It didn't last. Ted's pain increased again. Nerve blocks were unsuccessful. The doctors were losing their optimism. Further tests were made. The cancer was back, this time inoperable.

Ted, KT, Leah and me at 404 Cactus Drive   Photo: KT Budde-Jones

Dinkum the cockatoo shares the wheelhouse of the *Magruder* with me.
Photo: KT Budde-Jones

Despite this development, KT and I had to continue our work. With Dick gone I had no one to spell me as captain. First Mate Tony was good but not quite ready yet for full command. Our attrition-reduced crew had not been replenished due to the latest financial crisis, so every one of us was needed for the operation of the boat. The winter/spring weather was fortunately changeable, making our trips short so we could spend as much time with Ted and Leah as possible.

It was impossible to imagine life in Key West without Ted; he had been my boss, my great mentor and tutor. Over the years he had also become my best friend. Ted and Leah had offered both KT and I unquestioned support and love, we were inseparable. How would Leah deal with losing him? How would we? While trips to sea provided some diversion for us, Leah had none.

Mel and Deo were sympathetic as well, both being great fans of Ted and Leah. Mel however, planned a big early seasonal push on the *1715 Fleet* sites, which included using the *Magruder*. He wanted us to work more of the deep-water areas on the sites where the little local boats couldn't function. His intentions were for us to work the summer season there before returning to the *Atocha* and *Margarita* sites in the fall.

Once again we were pulled away from the *Atocha* and *Margarita*;

there was so much work to be done and it was frustrating to squander good summer weather somewhere else. This time it was even more distressing for us to leave as Ted was starting to fail rapidly. He had become housebound, and we felt the need to stay close.

Ted was critical of us acquiring Dinkum, but had quickly warmed up to the intelligent and endearing creature. When we brought Dinkum over to visit Ted, the avian strutted across the bed towards him. He quietly said, "I never thought that bird would outlive me." Fortunately, the extended Miguel family started to gather so there would be more help and support for Leah.

In mid-April we went back to Ft. Pierce and worked the sites as weather allowed. Fortunately, the weather was very changeable so when it got bad KT and I hired crewmembers John Hood and Maryann Zwierson to be "bird sitters" for Dinkum in our absence.

We often made the six hour drive down to the Keys for a day of visitation with Ted, and then drove back just in time for the weather's improvement. It was always a tedious trip, made stressful by having to balance the demands of the job with our need to spend time with the Miguel's. With years of ship deployments behind them, Ted and Leah understood perfectly.

One visit in early May we assumed duties of a bedside night shift while Ted hallucinated. The family had temporarily taken him back to the hospital since the cancer began affecting his mind. It was hard to see him so confused, many times during the night we had to restrain his movements. The next morning we had to return to Ft. Pierce.

We used what phone access opportunities we had between visitations to check on Ted's condition. He had calmed and was back home again in his own bed with the whole Miguel family at hand. We hoped for one last visit.

May 18, 1990.

Working a new area north of the Sebastian Inlet had produced several days of empty excavations. The early morning, low angled sun already blasted through the open watertight door and windows of the wheelhouse. We had just gotten underway, heading south towards another day of work on the *Cabin Wreck*. I filled out the rough logs, listing crew, event times and operation plans. The log was an important document. It was the ship's memory, reflecting for all time what we lived. I kept the logs much the same as Ted did years earlier.

"Crew: Syd Jones, Tony Absetz, KT Budde Jones, Dave Bass, Steve Hannon, John Hood, Maryann Zweirson. Weather: Sunny, seas one to two foot light chop, winds east 10 knots. Underway to *Cabin Wreck* at

0730 from night anchorage 2 miles ENE of Sebastian Inlet." I paused my writing as a radio transmission broke.

"Unit 21 *Magruder*, This is Unit One."

It was Taffi Fisher's voice. She ran the Sebastian museum and oversaw the recovery operations on the *1715* sites.

I picked up the mike. "Go ahead One, this is *Magruder*."

Her voice had an apologetic tone. "Syd, I just heard from Key West… Ted died this morning. Leah and his family were with him. It was peaceful. I thought you would like to know."

I could think of no profound response. " Thanks Unit One."

The open log lay on the binnacle shelf where I had placed it. I wrote in a new entry. "Ted Miguel died this morning in Key West."

Ted, the great salvage master, with his encyclopedic knowledge of almost any subject, my great teacher and friend was gone. In his usual calculated style, he had achieved financial security for himself and Leah, mellowing over the years to really begin enjoying life. Now there were no more pages in his book.

Ted was cremated and there was a service in Key West. The Treasure Salvors present and past alumni were there en mass. Many people, including KT and me, took turns speaking of their experiences with Ted. Easily the most profound of the remembrances came from Tom Ford, who perfectly expressed the reliance and bonds many of us operations people had with Ted. In his story he used Ted's self assigned radio unit, number 10.

"Over the years, every evening Ted would radio the boats out on the sites to make sure everything was OK. Everyone else at the office had gone home, but Ted would always call from his own radio station at home. It was so nice to know someone was thinking about us, that our presence out beyond the horizon hadn't been forgotten."

"When I was running the *Swordfish*, I'd look forward to that end of the day radio call. No matter how bad the day had been, I just couldn't help feeling better when I heard "WZG9605 Unit 10 to Unit 13, How's it going out there Tom?"

"Now, he's the one out beyond that horizon and it's my turn to return the gesture. Unit 10, Unit 10 this is Unit 13, can you read me? Unit 10, Unit 10… Unit 13 is standing by on this frequency for you."

Casa Key Haven

After the service we returned to Ft. Pierce and the *1715 Fleet* shipwrecks. With so many family members now living at the Miguel house, Leah had plenty of company. Some of them seemed to view our continued presence suspiciously, so it was time to leave. Getting back to the ship was good avoidance therapy for both KT and me.

The crew of the *Magruder* sidescanned, explored and excavated *Sandy*

Not long after Ted's funeral, Mel visited the *Magruder* in Fort Pierce. Here he plays with Dinkum the cockatoo. Photo: KT Budde-Jones

Point, Corrigan's, Cabin Wreck and *Colored Beach* to the best of our ability in the deepwater areas Mel had assigned us. On some sites we actually found numerous artifacts, on others just empty seafloor. It soon became clear that the deeper water areas were of limited scope and that the *Magruder's* oversized capabilities were being wasted. It just couldn't get into shallow enough water to approach the bulk of the remains. After I conferred with Mel, he decided that the *Magruder* should once again head back to the Keys and the *1622* sites.

We got underway on the night of July fourth. By staying inshore to avoid the Gulf Stream, were treated to a progression of firework displays as we passed each successive community. Most on board saw it as our own private celebration to be Keys bound.

Since Leah's family was staying at Cactus Drive, we spent little time with her when we arrived home. The *Magruder* quickly shifted back into the *Atocha* and *Margarita* recoveries. A recent addition to our normal operations was to video significant shipboard and underwater operations on every trip.

On a whim I decided to revisit the **Coral Plateau** gold coin area that years earlier had produced so many coins for the *Endeavor* and the rest of us. I wanted to see if our newer detectors and years of experience would make any difference in the well picked over spot.

That afternoon KT videoed me finding and recovering a two escudo gold coin among the soft corals. It had been covered over by a calcareous concretion. The coin turned out to be an extremely rare Santa Fe de Bogotá mint, one of the first gold coins made in the new world. All four known examples were found on the *Atocha*. The rest of the gold coins on the *Atocha* and *Margarita* were old world currency. The other divers did well also, finding silver coins and musket balls that had previously been missed. I felt pretty good about our abilities, and began revisiting older areas on both sites to see if our improved skills paid off again.

KT and I began visiting Leah more regularly as her family departed. We knew that she would face a big let down after their departure, and we wanted to be there for her sake as well as our own.

On one trip to her home we noticed an older house for sale just a few doors down the street. We had long ago written off ever living in Key Haven, anything on the water with a dock was just too expensive. We called the realtor anyway and found out the stats and price. The current owners had built a house upstate and were leaving the Keys. Although they wanted to sell quickly, their asking price was beyond what we could afford.

We made a low-ball offer, expecting to be turned down. Two days later we got a call. We were buying a house in Key Haven.

The planets really aligned. We put an ad in the newspaper to sell our townhouse. The one and only person who answered the ad bought it for our asking price. We were even able to take the furniture with us to Key Haven. Things were going way too smoothly.

In a month we closed on both properties and moved into 404 Cactus Drive. The home was located exactly half way between Mel and Deo's place and the Miguel's. How ironic that after all the years spent visiting and living at 417 Cactus, we would live just four houses away. Sadly, Ted would not be there to enjoy it.

Leah spent a lot of time over in our new digs whenever we were in port. We all got along as previously, but Ted's absence was palatable. It wasn't that we couldn't enjoy each other; it was that our relationship needed to evolve into a new dynamic. We all worked at it in our own way.

Adding to the sting, Leah was diagnosed with cancer. Although her operation was a success, the pain of living in Key Haven with a constant memory of Ted became unbearable to her. Here there was just too much connection with her past life, and our presence only made things worse.

She had always talked of returning to her West Coast roots and decided to take a rambling trip with her daughter Holly. They visited Oregon and fell in love with it. Soon she would move there, starting a new life far from Treasure Salvors and 417 Cactus Drive. The Miguel era in our lives was…over.

While the *Magruder* was having a broken transmission repaired on Stock Island, a short, pleasant man with a mustache summoned KT out of her engine room. Richard was the husband of the Jill London who owned Women for Sail. They were going to set up business operations in Key West and wanted to hire her "to watch over" their boat for a few months.

With the transmission fixed, we were determined to continue our work on the sites. Operational funds were thin again; Mel's continuing wildcatting projects had sucked up most of the money. I took over the lab work in port, leaving Tony to oversee repairs and maintenance on the *Magruder*. I had some idea of how to process the artifacts from many past discussions with lab personnel about their craft. My efforts were just a stopgap until a replacement conservator was found.

Sometimes our wreck site exploratory efforts proved very fruitful. I located a thin but well-defined trail of artifacts that led to **Fay's Cannon** that had been masked under the 1800's **Copper Clad** wreckage. The artifacts we found were a complete set of very well-preserved carpenter's tools and chests of spikes. There was even a large section of well-preserved *Atocha* structure - the first found outside the mud zone. The trail was long and straight enough to tie in perfectly with the **Fay's Cannon-Northern Anchors** line. Some of the tools still had the owner's marks on them. It was another unique find, which connected the two northern most features back to the trail leading from the **Emerald Cross**.

Working along the west-northwest mud trail from the small *Atocha* grapnel found by the *Swordfish* produced more artifacts. Exploration through this area was especially hard because the trail was often more hints than substance.

Satisfaction for us came when comparing our current abilities to those of the past. As I had previously decided, we sometimes revisited old areas that had been really hammered. One area I decided to test was the **Margarita Main Pile**. It had been overly excavated in the early years of the wreck's modern salvage. One day I set up the *Magruder* right over the area where the *Swordfish* had first dug.

Hundreds of excavations by the company boats and numerous subcontractors had been done at this same spot. The shallow sand had been reshuffled many times. It would be interesting to see if we could find anything. I was the only one in the company left who would recognize this bottom feature. After a quick blow from the mailboxes, I saw the shape of that first wonderful solution hole we rookie divers had worked the day we found the ***Margarita* Main Pile** - ten years earlier.

The *Magruder's* digging power had pretty much cleaned out the entire solution hole of sand but there were still several feet still in the center where the depression became quite deep. I ran my detector over the spot and immediately got a strong hit. Hand digging produced the remains of an old aluminum can. I put the detector back into the pit and carefully moved the loop back and forth with the machine now tuned to ragged edge sensitivity. In one corner a slight but repeatable variation of the tone was evident. I carefully hand fanned deeper and deeper and felt through the broken shell and sand fragments with my fingers.

Something small and hard was at the bottom of the narrow sand column. I pushed my pointed fingers against the sand's resistance so I could just grab it. I pulled out a small chunk of a gold bar, obviously chiseled off and used as "small change." Ten years ago I found my first gold in this hole, a link from a money chain. Deja vu indeed.

I took great pleasure in watching Tony carefully work out a small, delicate piece of gold jewelry he had found in a sea floor fissure. KT was always a detailed searcher and now the other divers were just as careful, finding objects in places that had previously been worked hard. It was impressive how far we had all come and our new finds made it seem sometimes like the wrecks were regenerating. We also ranged further afield in the *Margarita's* secondary scatter and found new groups of important artifacts.

Except for money issues and the occasional crew squabble, life working the sites was the most comfortable and least stressful it had ever been. I even had exclusive use of a company owned pickup truck. Despite the pleasant working conditions, Ted and Leah's absence left a big hole in our lives. Aside from our friendship, I had come to depend on Ted as a resource for evaluating my engineering, managerial or operational ideas. Now that all his knowledge was in the ether, I was on my own.

A final settlement had been reached with the principals of Jamestown Treasure Salvors, and Mel regained ownership of the *Swordfish* and *Dauntless*. Both boats were now wrecks. They had continued to degrade even further from the way Jamestown had left them in the boat yard.

Kane Fisher had always loved the *Dauntless*, and now that it was back in his father's control he wanted to rebuild it and captain it once more. The *Swordfish* was actually in better shape, but since there was barely enough money to fix up one of them, Kane got to revive his favorite. The *Swordfish* would be scrapped.

Another group of subcontractor hopefuls hooked up with Mel and tried to fund their own proposed project out on the sites. They acquired a barely floating 1920's steel motor yacht and hoped to turn it into the ultimate treasure-hunting machine. It was a family affair; everyone associated with the operation seemed to be related. Typical of the uninitiated, they brimmed with new ideas of how to equip their boat so as to outperform anything we had ever done. One of their favorites was to install an inverted periscope on the vessel so that they could see the wrecks from the comfort of the wheelhouse.

Their boat's condition was scary. The hull was a rusty colander with many holes in it just above the waterline and the interior structure was largely rotted away. The old cabins and compartments were in complete decay and the engine room was a rusty cave. Concrete had been crudely poured over the back deck to repair all the rust on it.

Natural selection quickly ended their project, as potential investors were not impressed. Eventually the now abandoned yacht was towed sea to be made into an artificial reef.

Women for Sail approached KT. They wanted to hire her for a few months as an instructor for their sailing courses. I was a little nervous about losing a good engineer, but she assured me that Tony's work with her gave him the ability to fill in. Besides, it would be a great opportunity for her since some of their trips would be in the Mediterranean.

As the New Year sputtered in, company finances became an increasing problem. The rebuilding of the *Dauntless*, Mel's various wildcat projects, and both the *1715* and *1622* fleet operations all sucked up money faster than it came in. There were much longer intervals between our trips, and it wasn't always due to bad weather. We continued to lose crewmembers to boredom or uncertain paychecks.

New replacement diver Geoff Zitver came with more experience than most. On his own he had successfully worked some of the Spanish *1733 Fleet* wrecks sunk off the middle Keys. Other newbies had more normal backgrounds; Don was an ex-postman who took up commercial diving and Greg had recently graduated from college. Ed Hinkle returned from a sabbatical.

Kane soon brought the reincarnated *Dauntless* out to the sites. Its presence was somewhat short term, as he soon took the vessel and crew off on other adventures unknown to me.

With KT gone on Women for Sail trips, I was very busy. I ran the boat and tried to keep Dinkum entertained and out of trouble. It was like having a three-year old child on board. We had a kind of birdie playpen for him on the large back deck storage chest where he could keep an eye on us and we on him. The box was a convenient place for divers to temporarily put their finds while they took off their gear. Dinkum took more than one opportunity to try to grab up an emerald in his beak when he thought no one was looking. Tony helped me greatly by filling in as engineer so I could keep an eye on the feathered fireball.

Just as KT was ready to come back, Mel officially ran out of money. Again. For the foreseeable future there was no money for paychecks or for running the boat.

As had happened to their predecessors many times before, my crew would each quickly have to find their own financial life ring. Some found new jobs locally. Others disappeared for good. Poor Tony, who had finally proved himself ready for full command, decided that he had experienced enough of Mel's chronic insolvency. KT scrambled to get rehired by Women for Sail, and was lucky when they took her back. Mel allowed me to keep one employee to help me maintain the boat and Geoff Zitver stepped forward with an obvious eye on Tony's vacated position.

After many years of these cycles this process was so, so familiar to me. I expected things would eventually work out as they always had and I'd be trying to hire back my crew in a few months. There was so much left to do on the *1622 Fleet* sites…

NOVEMBER 1997

I made a final check on the straps that secured the new Lotus to the trailer. Funny, I still thought of the car as new even though KT had bought it for me almost seven years ago during Salvors' last big layoff. We were just about loaded up now and the moving van would be leaving soon. It was time to get the birds into the pickup.

These last years had been such a blur, never as intense in feeling as the first experiences at Treasure Salvors, but just as much of a roller coaster. For now we both needed to get off the ride, if only for a while. Our official leave of absence was about to start.

Our lives had jinked in unexpected ways. I had become Mel's operations manager and head conservator. KT was the education director for the Museum, which had become the most visited historic attraction in the southeastern United States. She also taught marine archaeology at the local college, updated her popular research booklet on Spanish colonial currency and she sang opera in the local community chorus. Together we had helped to establish a local wildlife rescue organization.

The last big layoff had forced us both to find new niches. With no money to run the boat, I spent more time in the lab processing all of the waiting artifacts. Just as Ted would have approached the problem, I spent a great deal of time studying reference and chemistry books to establish the best possible way to deal with different artifact types. I created a database to inventory and control the many thousands of items now in my completely reorganized storage and processing areas.

November 1997

The *Magruder* was kept operational in the expectation that we would go back to work, but I found the new challenge of preserving all those hard-won artifacts very interesting. I enjoyed the research and learning process and tried to make the lab better than it had ever been.

Deo Fisher became a great advocate for my plans and helped me greatly in acquiring equipment. Her public persona was far more demure than Mel's showman style, and I came to appreciate her strength of character, personality and intelligence. She had passed those same traits on to her daughter Taffi as well.

I had a hard time coaxing money out of Mel for any of the *1622* project's needs. All our remote sensing and survey equipment was now obsolete junk, and the *Magruder* badly needed boat yard repairs. He was currently having the time of his life engaged in his latest Civil War wildcat project. Deo always managed to come up with some resources so that at least my lab projects could proceed. I spent many enjoyable hours talking to her.

KT spent two more years working for Women for Sail. She sailed worldwide and became an expert captain in her own right. When being away from home started to get old, KT applied her Masters in Education qualifications to the Museum. She developed after-school programs, archaeological seminars for adults and later taught associative courses at the local community college. KT spent much time at the Museum, the college or Mel's corporate offices to evaluate and grade coins for Deo.

The company's finances kept on wobbling, so as a reserve I took a weekend job running a 70' ketch for snorkel and sailing excursions. The beautiful yacht was docked at the Hyatt hotel and was operated through the same family that owned Women for Sail. I appreciated the extra money and as the only crew on board; it kept my maritime skills sharp.

The biggest challenge to Salvors' continued existence began during plans to make all the Florida Keys into a national marine sanctuary. Suddenly the company rumor mill went white hot as proposed or suspected intents by the Sanctuary's managing bureaucracy for "submerged cultural resources" evolved. Mel, Deo and company had already experienced years of nasty court battles against Federal and State governments over shipwreck ownership rights. From the company's perspective this had all the overtones of yet another government property grab.

Mel's fight with the Sanctuary officials quickly went from just words to actions. Under the ill-advised guidance of a Czechoslovakian national (who had several times previously tried to pirate the *Atocha* from us), Kane took the *Dauntless* to the *1733 Fleet* wreck sites off the middle Keys. Not far offshore from where the new Sanctuary's offices had been established, he began looking for wreck material by digging holes in an area known as **Coffin's Patch**.

Many artifacts found during the operation were delivered to my lab for processing. By now I had established standardized protocols for recording and stabilizing the finds. My new lab's techniques would soon be put to the test.

The Sanctuary officials took helicopter rides out to the site to film the excavations visible in the shallow water and send divers down to photograph and measure the sizes of the excavations. These officials were interviewed in the media and declared that the damage done to the seafloor and the provenience of the site was irreparable. Charges were filed to protect the environment and historic sites now within the proposed Sanctuary boundaries. The maps they produced showed the sanctuary's control area also included both the *Atocha* and *Margarita* sites.

KT and I were involved with another type of challenge during our off times. In our own, often-surveyed patch west of the Marquesas Keys, species after species was disappearing from over fishing. Once the fishermen reduced one animal group past economical harvesting they simply switched to another. We had watched fisheries of swordfish, kingfish, shrimp, grouper and lobster decimated or extinguished completely. Birds and other creatures also suffered from human activities. In our spare time we and Becky Barron established Wildlife Rescue of the Florida Keys.

Starting up the new enterprise amidst our swirl of work activities was exhausting but rewarding. Eventually we negotiated with the city for a site for the organization, built cages and pens, did fundraisers, trained volunteers and took classes in wildlife rehabilitation.

My lab project steadily progressed. The drawings, film photographs, and measurements made before and after processing eventually phased into digital photographs integrated into the artifact records. John Cochran, an ex-student of KT's college course began working for the company and used his drawing skills to help do onsite documentation of the finds. I took over representation of the company's conservation activities before the Admiralty Court.

Federal lawyers soon deposed me over the pending Sanctuary lawsuit against Mel. They asked me to produce my lab records for the *1733* artifacts that Kane had recovered. By this time, all the items were cleaned, stabilized and stored in an environmentally controlled area. Hours of questions about my processes had to be answered, with documentation to backup my claims. Their conservation experts reviewed my statements and examined the artifacts. I felt the cross hairs on my back. They looked for anything to use against us.

When just enough money had become available to take care of the *Magruder's* now serious boatyard need, KT and I delivered the ship to an inland Gulf coast boatyard where the yard workers would perform the rebuild.

Because I was absorbed now with so much other critical company business, Mel and Deo agreed that I should assume the position of operations manager and hire a new captain and crew for the *Magruder* when it was ready. I found a charter boat captain who was interested in the job and trained him and his new divers when the boat was back in the water. Mel then sent them and the *Bookmaker* on a survey project for wrecks near the panhandle of Florida - a long way from the current battle with the Sanctuary.

Many media salvos were exchanged for and against the sanctuary concept. Pat Clyne took up Mel's cause and even set up a television studio in his office, producing a weekly anti-sanctuary program broadcast over a local cable access channel. There were frothing exchanges in the local editorial pages, demonstrations and vocal public hearings. The anxiety factor around the office was pegged out.

With the panhandle project finished, the *Magruder* and the *Bookmaker* returned to the Keys. The crews made tentative trips out to the sites, but the mailboxes were not used. No one was sure what the next move would be.

Eventually the Florida Keys National Marine Sanctuary was established by popular vote. Tensions slowly eased as both sides began to explore their mutual options. I took one of the Sanctuary representatives to the sites and showed him the various seafloor features along both wreck trails. After analysis, they eventually agreed that our excavations in the mud or sandy areas had no real impact on the environment.

The captain I had hired for the *Magruder* quit, treasure hunting just wasn't in his blood. His replacement didn't last long either. Though verbally enthusiastic, he spent far too much time tied to the dock in good weather. I was forced to dismiss him.

It was time for me to start considering the option of my having to run the boat again. How exactly was I going to do that with everything else that was expected of me? Fortunately, the answer to my problem came from Gary Randolf, an eager beaver diver whom I had hired earlier.

Gary was smart, motivated and trainable. Years ago I would have considered him too inexperienced for the job, but there was no longer a pool of established candidates from which to pick. I made familiarization trips to show him how to maneuver the boat, how to recognize the various wreck site features and how setup operations should work. We spent a great deal of time in my office going over site cartography. At last he felt ready for the challenge and perhaps much like my own beginnings, more luck than skill saved the day. Fortunately for both of us he was a quick study.

With our old Del Norte system now on its last gasp, we had to come up with a quick and viable alternative for site navigation. Mel had no money left to purchase any new Del Norte equipment so we decided the cheap and practical way was to use Differential GPS. It wasn't as precise as the Del Norte had been, but it wasn't as bad as LORAN either. I made a final trip together with Gary using both the Del Norte and new GPS, taking readings at many benchmarks along both sites. In port I made site map overlays that would allow the use of the old plotted data with the new format. Gary learned the technique as well and soon made his own charts.

Although we had been cleared by Sanctuary officials to use airlifts on the *Atocha* site, they still hadn't yet OK'd mailbox use. The boat crews were happy to at least have some options, and began dredging for emeralds much as Kane's barge crew had done years earlier.

The *Dauntless* eventually returned to the site but never again with Kane. Mel and I gave command to First Mate Robbie Hanna. The *Swordfish* had been stripped of its engines and equipment and was sunk as an artificial reef off Boynton Beach. Bob Kunke, who performed much of the dismantling process, gave me a brass porthole and ship's wheel as a keepsake. He gave the compass binnacle to Tom Ford.

Though the *Atocha* and *Margarita* sites were very slowly returning to our control, Mel ultimately lost the court battle over Kane's excavation on the *1733* site. My own hard work was personally vindicated in the final decision; the government found absolutely no fault with any of my lab methodology. Their main case focused on the unauthorized damaging of protected turtle grass beds in the Sanctuary. A large fine was levied. All the artifacts were to be surrendered to NOAA Sanctuary officials.

During the windup to the trial I proposed, through our attorney, that the artifacts in question be donated to the non-profit Mel Fisher Maritime Heritage Society's museum. This wasn't a joke - the museum had a huge following and was now quite independent of Mel's control. If the new Sanctuary administrators really felt the best use of the objects was permanent public display in the region from which they came, this was actually the best option. Not surprisingly the idea was rejected.

I got in one last good tweak on the bureaucrats. After the trial, I was notified by phone that a representative would be coming to inventory and collect the *1733* artifacts from the lab. I agreed to the meeting and explained that I would have paperwork ready for him to sign when he took the items, declaring that the objects were in his care and he would assume complete responsibility for any future damage or degradation when they left my control.

This little statement apparently caused a giant tail chase up and down the bureaucratic pyramid - no one wanted to personally sign their name to that responsibility. Later, government officials decided that the items should be donated to the Mel Fisher Maritime Heritage Society for public display.

Gary needed some vacation time so I spelled him for a trip as captain

November 1997

After all the millions of dollars in Spanish treasure it recovered on both *Atocha* and *Margarita* sites, the *Swordfish* was stripped and sunk as an artificial reef. This wheel and porthole are some of her few surviving artifacts.

on the *Magruder*. I did not know it at the time but it would be my last visit to the *Atocha* and *Margarita*. By now the Sanctuary officials had granted us limited use of the mailboxes so it was almost like the old days. Dick, KT, Tony and all the others with whom I was so familiar had been replaced with new faces.

It was an unremarkable trip. I spent a lot of time in my own thoughts, sitting on the back deck while divers worked below in the latest excavation. To the east, the green topped Marquesas Keys floated like a mirage just above the horizon.

So many times the sight of those islands had drawn me into thoughts about Melian, Vargas and the other Spanish salvors who preceded us. The view from my boat was exactly what they had seen, and our triumphs and failures often mirrored their own experiences hundreds of years ago. Sometimes when we found their lost salvage equipment it was if we could shake hands. Their world seemed so real to me.

And so it should. Admiral Moran, so anxious to search for the Capitana and Almiranta. Wasn't he like Bob Moran? Vargas, the respected old salvage master who's vast knowledge exceeded his luck in finding the lost ships. Was he not Ted? Melian, the great self-promoter, a man of schemes and dreams hoping to leverage his finding of the lost ships into a position of note and recognition. He would fail to achieve finding all of the missing treasure, and would make plans until his death for continued attempts. Above all, what he did find earned him the social position he craved. Who else could this be other than Mel himself? The *Atocha* had finally granted him his favorite title "The World's Greatest Treasure Hunter". Even Key West titled him "The King of the Conch Republic". He was known on every street and bar on the island.

The value of the gold, silver, emeralds and artifacts that we had recovered on the *1622 fleet* sites was estimated to exceed 500 million dollars. Mel parlayed his great find into money that allowed his imagination to run wild. He looked for Atlantis. Could things have turned out better? He would always imagine that maybe next season we would find the rest of the missing *Atocha's* treasure. And just like Melian and Vargas, we would never find all of it.

Mel often introduced me as his "Right-hand man". Especially due to our completely different perspectives of the world, it was a wonderful compliment. Ted would have smiled.

The moving van pulled away. Two doves, both Wildlife Rescue refugees that couldn't be released, joined KT and Dinkum in the Land Rover. I drove the pickup and towed the trailer with the Lotus on it. We had sold our house at 404 Cactus Drive. Hurricane Andrew had convinced us that our ground level house might be a poor choice when a major storm finally hit. We already had a newer stilt house picked out in another part of Key Haven. The realtor was waiting for our answer. For now, we needed a break. We put our things in storage and wondered what new adventures we might find hanging out at an aircraft restoration museum in Kissimmee. Our leave of absence granted us about two months to find out.

Epilogue

We never moved back to Key West. We traded our familiar and comfortable lives for the promise of new adventure working at the Warbird Museum. In many ways it proved to be as profound, confusing and frenzied an experience as Treasure Salvors had been. KT and I flew our own WWII fighter trainer, I was type-rated in B-25 bombers and we flew the air show circuit together. As at Treasure Salvors, our jobs broadened quickly.

We were next hired to help start a new aviation museum in Pearl Harbor. KT and I found and recovered the only surviving wreckage of a crashed Japanese Zero shot down during the Pearl Harbor attack in 1941; another of the never ending spin-offs of our association with Mel, Deo and Treasure Salvors. There have been far too many to count. We met and were befriended by so many fascinating veterans, both nameless and famous.

After years of fighting cancer, Mel died about a year after we left Key West. There was a wonderful remembrance ceremony in the courtyard of the Museum. Mel's old friend Cliff Robertson spoke about his old adventures with Mel, and KT sang Amazing Grace. The courtyard was packed with people, all happy to see each other one more time and recount old stories. At the end of the night we all danced and sang one of Mel's favorite songs, the Hokey Pokey - just like we had at so many Fisher Christmas parties years earlier. Mel was right… that's what it's all about.

Gary and his wife Linda painlessly replaced KT and me in the company hierarchy. Kim Fisher married Lee and after Mel's death assumed the corporate reins. He did a great job of changing the company's financial situation to a consistently more positive status. Kim focused with Gary on upgrading the company boats and equipment and they redoubled their efforts on the *1622* sites. The initial distrust between the Sanctuary officials and the company eased further with new management on both sides. Though some nice finds have been made over the years, the 8 missing cannons, 200 silver bars, thousands of coins and the rest of the manifested gold of the *Atocha* still eluded discovery. What remains of the *Margarita's* cargo has been even more elusive.

After her husband's death Deo, like Leah, sold her house in Key Haven and moved near Vero Beach to be near her daughter Taffi. She died in 2009 of cancer. KT would begin her own battle with cancer in 2010.

The boats still go out as long as there's good weather and money. Dinkum's chew marks still mar the wood in the bridge of the *Magruder*. Young divers unknown by me still push detectors against the current looking for long lost artifacts.

Unit 10, Unit 21 is still standing by for you on this frequency…

Want to know more about the treasure coins of the Spanish Plate Fleets?

This informative 28-page booklet, authored by KT Budde-Jones, contains maps, diagrams and listings needed to identify the cobs and coins of the Spanish Realm found aboard the Plate Fleet vessels. Assay listings from 1535 - 1824 are accompanied by clear, concise charts of the obverse and reverse coin impressions. Containing photos of escudos and reales, this is a publication of real value to collectors, and laymen alike. It is designed as an accessory document for anyone buying or selling treasure coins.

To order, contact:
KT Budde-Jones
321 437-9666 or
captkt@juno.com

INDEX

1588, 101
1600, 101
1616, 103, 193
1620, 103
1622, 20, 192, 199
1627, 188
1629, 188
1715, 20
1935, 15
1975, 67
1976, 8
1979, 30, 88, 94
1980, 109, 126,
1981, 187
1982, 216
1985, 276
1986, 280
1987, 282
1305 lbs of copper, 129
1622 Fleet bullion, 220
1622 Fleet sites, 323, 224
1622 galleons, 167
1622 Tierra Firma fleet, 103
1622 wreck sites, 294
1622 wrecks, 307
165 degree line, 270, 269, 272, 274, 292
1700's English merchantman, 300
1715 Fleet, 111, 205, 222, 295
1715 Fleet sites, 207, 279, 314
1715 Plate Fleet, 137, 197
1715 silver coin, 139
1715 sites, 144, 310, 316, 317
1733 Fleet wrecks, 322, 325, 326, 328
17th century swivel gun, 237
1975 deaths, 89
1980, July 6th, 148
1981 summer, 188
1983 New Year, 218
1984, January, 221, 222
1985, July 19, 239
1985, May, 229
1985, Memorial Day, 232
1985, October, 268
1986 New Year, 268
1997, November, 324

200 silver bars, 332
20th century discards, 97
404 Cactus Drive, 330
417 Cactus Drive, 313, 320
425 Caroline Street, 170
47 tons of *Atocha* silver, 256
500 million dollars, 330
8 missing cannons, 332

A

a low rise on the sea floor, 239
A&B Lobster House, 44, 48
abrupt plunge, 97
Absetz, Tony, 296, 302, 308, 315
Admiral, 40
Admiral Moran, 330
Admiralty claim, 133, 303
Admiralty Court, 294, 326
Admiralty Law, 118, 189
aerial guidance, 166
aged timber, 235
air conditioner drippings, 232
Air Force, 49
airlift, 177, 192, 193, 235, 265
Alice, 175, 177, 181, 187, 196
Almiranta, 103
aluminum cans, 76
alumni, 316
amount of treasure, 162
anchor, 179, 192
anchor lines, 154
anchor, large, 97, 193
anchors, 61, 118, 193
anomaly, 64, 65
antennae cable, 82
Antigua, 298, 299
April, 315,
Aquapulse, 198,
Arbutus, 54-56, 79-84, 88, 95,
113-123, 161, 188, 195, 208

Arbutus crew, 172
Arbutus duty, 201
archaeology crew, 242

Archive of the Indies, 90
Arden, Jo, 195
arquebus, 147, 182, 150
arrest warrants, 268,
artifact highway, 269,
artifact tags, 243,
artifacts, 119, 124, 133, 249, 302
Atlanta Constitution, 236,
Atocha, 7, 20, 54, 63-76, 102, 106, 147, 159, 199, 224, 318

Atocha and *Margarita* sites, 326
Atocha artifacts, 61
Atocha corridor, 117, 202
Atocha Main Pile, 240, 242, 269-275, 302-305

Atocha project, 10
Atocha site, 35, 42, 55, 85, 135, 143, 204
Atocha trail, 166
Atocha, salvage of the , 99
Atocha, scattered artifacts, 67
Atocha's "main pile", 153
Atocha's protruding mizzenmast, 107
Atocha's wreckage, 50, 216
August 27th, 167
Australia, 220, 222
Averia, 103

B

Bahia Honda, 16
ballast, 76, 184, 275
ballast dump, 249
ballast stone, 119, 237, 249
ballast stones, 73, 75, 127, 260
Bambi, 33
Band-Aid box, 170
BANG!!!, 82
Bank of Spain, 98, 110, 203-206, 213, 216, 254, 260

Banon, Juan, 121
bar of silver, 21
Barbuda, 298, 300
barge, 289
Baroque Dolphin, 24, 25
barracudas, 62
Barron, Becky, 326
Barron, Bill, 251
Bass, Dave, 315
Beaufort, 304
Beckman, Larry, 122, 145, 186, 217

beer can tab, 76
bell, 121
Benchley, Peter, 164
bends, 253
bent, 297
Bernie, 201
Bettencourt, J.J., 241
beyond the horizon, 316
big anchors, 60
Big Guys, 290
bilge pump, 82
Bill the bookkeeper, 35
black rocks, 76
Bleth, 220
Boat Bar, 46, 47, 158
Boca Grande Channel, 289
bomb, 67, 130, 231, 241
bomb crater, 75
bombing range, 67
bookkeeping, 220
Bookmaker, 159, 190, 202, 206, 218, 226, 228, 235, 303-307, 327

bored, 64
Borum, J.C., 94, 179
Boston Whaler, 50
bow anchor, 143
bower, 192
Boyd, Craig, 113
Boynton Beach, 328
Brady Bunch, 278
Brandon, John, 202, 205, 310
brass flare pistol, 47
Breese, Denny, 164, 175, 176, 207
broken pottery, 76
Bronze Cannon, 19, 147, 182, 224
Bronze Cannons, 182, 263
bronze Madonna , 180
bronze mortars, 151
brownish particles, 38
Budd, Ralph, 256
Budde, Katie, 175
Buen Jesus, 106
Buffet, Jimmy, 44, 247
bulk of the remains, 318
Bull and Whistle Bar, 170, 171
buoy block, 192
buoy line, 77
bureaucracy, 325
burned out, 208
Busch, Paul, 242, 251, 267-268
Butler, Dennis, 186

C

Cabin Wreck, 223, 310, 315, 318
calcareous concretion, 205
camera crews, 246
Candella, Steve, 284
Cannon, 162
Cannonball Clump, 153, 157-163, 172, 179, 182-184, 198

cannon carriage axle, 248
cannonballs, 150, 207
Cannons, 113, 118, 176
Capitana and *Almiranta*, 330
Capt. Danny, 72, 77, 78
Capt. Tony's Saloon, 30
Captain Ed, 230, 232, 235,
Captain Vargas, 106
captain's license, 221, 289
Carlos, Juan, 167
Carrera de Indies, 102
Cartagena, 102, 104
Cash, Jerry, 176
Castilian, 109-122, 130, 183
Cateraita, Marquis de, 3
Cayos del Marquis, 1, 54, 107, 115
Central America, 303-307
Central America fantasy, 311
ceramics, 159, 183
chain shot, 237
challenge, 158
Charleston, 305, 309
chart board, 92
chart recorder, 64
Chaves, Juan de, 169
cheerleaders, 292
Christmas, 293
Christmas arrived, 177
Christmas recess, 175
chronic insolvency, 323
cinta, 205, 207, 216
circle search, 65, 150,
Civil War era wreck, 119
Civil War wildcat project, 325
Civil War wreck, 98, 223,
Clem, Mike, 179
Clifford, Barry, 224
clump of blackened coins, 125, 128,
Clyne, Pat, 135, 149, 150, 204, 220, 227, 256, 306, 327

Coast Guard, 88, 207, 297
coastal survey, 223
Cobb Coin Inc., 295
Cochran, John, 326
cockroaches, 85
code number, 95
Coffin's Patch, 325
coin chests, 251
coin credit, 95
coins, 120, 139
coins and bits of gold jewelry, 311
college, 29
Colonel Green, 174
Colored Beach, 223, 318
Columbus, 100
commercial diver, 157
computer program, 275
conch, 198
concrete block buoys, 65
concretions of silver sulfide, 250
Confederate blockade-runner, 307
contract archaeologists, 254
cooking pot, 118, 159
coordinates, 64, 167, 237, 304
Copper Clad, 98, 320
copper ingots, 124, 127, 239, 241, 255, 262, 263, 275

Coral Plateau, 98, 204-209, 216, 225, 319
Coral Plateau gold bar, 211
Corrigan's, 223, 318
coup-de-tat, 227
Cousteau, Jacques, 6, 137
Craig, 92, 93
cranky, 162,
crap in the water, 37
crazy Englishwoman, 300
Cream rises to the top, 93
creepy noises, 84
Crew, 46
crew boat, 110
crushed ice, 52
crushed wooden boxes, 250
Cuba, 58, 189
Cuban refugees, 133,
Culpepper, Sherry, 279
cutting the lines, 79
Czechoslovakian, 325
dagger, 221

D

Daley, Bob, 114
Danny, 72-79

Dauntless, 172, 180, 188, 194, 239
240-241, 253, 265, 321-328

dead accurate, 249
dead rats, 37
dead reckoned, 304
deaths, 7
Debbie, 194, 207
debt, 47
December, 86
decompression chamber, 297, 232
Deep Bay, 300
deep-water areas, 314
deflector, 50
Deja vu, 321
Del Norte equipment, 182
Del Norte mapping, 216
Del Norte system, 327
Del Norte system began failing, 302
Del Norte tower, 209
dense mud layer, 97
Deo, 332
derelict vessel, 81
DeSchmitt, Jacques, 302
detailed mapping, 131
detector, 192, 233
detector hits, 188
detector searches, 204
detector survey, 210
Dick, 304, 311, 329
Differential GPS, 328
digging, 74
Dinkum, 313-322
dirtbag derelicts, 42
dirtbags, 49, 95
discharge hoses, 36
discovery of a lifetime, 240
Disney World, 15
dive tanks, 51
Diver's Alert Network, 298
diving accident, 231
diving bell, 183
division, 220, 277, 280
division bonus, 279
division program, 278
dogs, 80
dolphin, 25
dominant force, 17
Domingez, 98
Dominicans, 230
Donnie, 284
Door County, 7
Dorwin, John, 270

Dr. Lyon, 109, 127, 183, 267
dragging anchor, 147
Dreams of Gold, 278
dredge crew, 116, 160, 161
dredge project, 195
dredging for emeralds, 328
drifters, 49
drinking, 47
drinking water, 83
drug smuggling, 198
drunken antagonisms, 48
drunks, 30
Dry Tortugas, 53, 106, 107, 154
Duncan, 254
Dunn, Phil, 302
Dunn, Sandy, 302
Durant, Don, 94, 122, 131, 137, 230, 240, 275
Dutch, 104
Dutch pirates, 188
Dutch privateer, 178
Duval and Eaton, 46
Duval Street, 30

E

E.O., 73
East Martello, 199
Easter Island statue, 22
economic frailty, 89
Education Director, 324
eight bronze cannons, 263
eighty-pound bars, 242
electrolysis tank, 302
electronically survey, 50
embolise, 253
Emerald City, 262-275, 289
Emerald Cross, 260, 320
Emerald fever, 265
emerald jewelry, 234
emeralds, 311
Endeavor, 202-207, 223
envelope, 303
Etchman, Bruce, 176
ex-Coast Guard buoy tender, 55
ex-shrimpers, 49

F

fantasy trip, 133
Farr, Kevin, 94, 123
fatal consequences, 45
fathometers, 53, 58, 91

Fay's Cannon, 224, 260, 320
Fay's Cannon-Northern Anchors line, 320
Federal Admiralty, 86
Federal and State governments, 325
Feild, Fay, 114, 160, 173, 176, 222, 224, 279, 302

fell overboard, 58
Ferrera, Captain Alonzo, 103
Fiat, 28, 181
Fiddler in the Bilge, 202
financial toilet, 293
fire sale, 286
firearms, 289
firearms on board, 88
first mate, 289
First Mate John, 72
first name only kind of town, 29
Fisher Factor, 284
Fisher, Angel, 71, 195, 242
Fisher, Delores, 39, 195, 221, 325
Fisher, Dirk, 71, 98, 147, 242, 254
Fisher, Jo, 228
Fisher, Kane, 289, 321
Fisher, Kim, 137, 190, 202, 228, 331
Fisher, Taffi, 276, 278, 280, 316
Flagler, Henry, 15
flare gun, 47
Fleet Silver masters, 104
flight training, 12
flim-flam man, 95
Florida and the Federal government, 89
Florida Bay, 17
Florida Keys, 7, 105
Florida Keys National Marine Sanctuary, 327
Florida turnpike, 13
flying lessons, 181
Ford, Tom, 135, 216, 218, 220, 225, 237-239, 263-269, 277, 283, 293, 297, 301, 316, 328

Ford, Tom fortuitous mistake, 236
Ford, Tom replaced me as Captain, 226
forty-seven tons, 160
found the bulk, 282
four overlays, 260
fourth of July, 189
Frank, 157
Frank attacked , 158
Fred, 27-48, 70, 72, 77-79, 88, 92
Frick, Olin, 86, 88, 127, 189
Front Street, 18, 30, 47
frustration, 194

Ft. Lauderdale, 181
Ft. Pierce, 202, 207, 307, 310, 315, 317
full-blown gale, 307
funds were thin, 320
futtocks, 262

G

Gage, Rick, 71, 242
gale, 82
Galleon, 18-36, 45, 49, 112, 129, 130
Galleon Anchor, 97, 110, 118, 202-203, 215, 260

Galleon museum visitors, 70
Galleon repair diving, 45
Galleon sinking, 36
Galleon sinks, 135
Galleon was burned, 207
Galleon, perpetually leaking, 88
Galleon's bilge pumps, 38
galleyware, 274
Garrison Bight Bridge, 17
Gary, 328, 331
Gasque, John, 86, 88
Gasque, Rick, 127
gay community, 198,
ghosts of the *Atocha*, 84
girl friend, 187
glorious spot, 151
gold and emerald jewelry, 233
gold and emerald ring, 205
gold bar, 123, 216, 267, 321
gold bars, 120, 133, 162, 163, 242, 263
Gold bullion, 161, 229
gold bullion found, 211
gold bullion markings, 220
gold chain, 162
gold clasp, 213
gold coin, 221
gold coins, 161, 207, 259
gold cross, 215
gold crucifix, 258
gold fever, 133
gold jewelry, 233, 269, 272
gold money chain, 172
gold money chains, 150
gold plate, 161
gold ring with an emerald, 128,
gold rush, 222
gold two escudo coins, 199,
Golden Doubloon, 20
Golden Venture, 201-208, 211-216

good luck, 25
Good Morning America, 270
governments , 190,
Governor of Havana, 135, 264
gradiometer, 113,
grapnel hook, 263, 320
grapnel, smaller broken , 178
Graveyard of the Atlantic, 305
Grayden, 303-311,
Grayden's folly, 306,
great mentor, 314
Green Cabin, 223
green cigarette smoke, 155
green flash, 114, 115
Green, Alice, 174, 218
Green, John, 95, 117, 127, 157, 162, 163, 194, 220

Greene Street, 292
Greg, 291, 322
grid system, 250
grid system unpopular, 251
Guard Fleet, 102
Guard galleon, 103
guardian Jesuit priests, 293
guests, 170
gilded silver cups, 269
Gulf of Mexico, 14, 106
Gulf Stream, 17, 103, 305, 318
gun, 47
gunshots fired, 88

H

Halfshell Rawbar, 312
hammer code, 261
hand held blower, 130
hangers-on, 249
Hanna, Robbie, 328
Hannon, Steve, 315
Happy Birthday, 261
Hargraves, T.T., 195, 220
hatch covers, 80
haunted, 84
Havana, 103
head, 38
Heather, 312
Heavy ballast, 151
heightened workload, 263
Hemmingway, 30
hemp rope, 262
Here we go..., 65
High Mass, 105

high quality gold, 128
hit, 65-73, 96, 115, 117, 176, 239
Hokey Pokey, 331
holidays, 281
homeless dirtbags, 198
Homestead, 14
honeymoon, 120
Hood, John, 315
Horan, David Paul, 118, 306, 311
horror stories, 55
hot shower, 85
Hot Tub, 172, 173
Houseboat Row, 89
huge emerald, 215
huge waves, 82,
hull spikes, 207, 288
human bite marks, 180
hundreds of coins, 310
hurricane, 105, 184
Hurricane Andrew, 330
Hurricane Bob, 255
Hydraflow, 55
hydraulic cement, 36
hyperbaric chamber, 297

I

impact scars, 274,
in the middle of it, 142
inaccurately recorded, 96
inflatable boat, leaky, 79
Ingerson, Rick, 179
intact ceramic jar, 269
intact olive jar, 242
intact wood, 127
inventory all of the artifacts, 302
inverted periscope, 322
IOUs, 220
iron bars, 117
iron cannonballs, 63
IRS approved appraisers, 286
ivory box lid, 164

J

J.B. Magruder, 225, 246, 247, 260, 280
James Bay, 93
Jamestown Treasure Salvors, 283-294, 321, 302

Jeff, 29, 35, 42-49, 55, 70-79, 88
Jefferson, 41-46, 50-55, 61-88

Jefferson's excavations, 69
jewelry, 161, 162, 275
Jimenez, Martin, 274
Joan, 13
Joe, 27, 37, 66, 73
John, 30, 76, 203, 204
John Pennekamp State Underwater Park, 15
Jonas, Donnie, 225, 260, 280, 283, 246,
Jones, Syd, 186
Jordan, Bobby, 109, 129, 190, 217-218
July fourth, 318
Jumper, 86
June 17th, 140
Juniper, 86, 88
Juniper's crew, 87

K

Kane, 35, 113-126, 136-137, 159, 163, 172, 197, 208, 213, 220, 224, 241-259, 328

Kane was discharged, 291
Kane's Trail, 229
Karen, 220
Kevin, 299
Key Haven, 89, 319, 332
Key Largo, 15, 17
Key West, 8, 15
Key West Welding, 284, 284
Key West's favorite sons, 129
Kim, 197
Kincaid, Don, 114, 119, 132, 149, 150, 200, 220, 255

King of Spain, 167
Kirby, Sam, 293
Klaudt, Dick, 161, 176, 182, 208-218, 226, 277, 283, 287

Klein, Bobby, 189, 190, 194
knives, 269
Koblick, Ian, 201
Kopp, Tony, 298, 299
KT, 175, 177, 187-331
KT Education Director, 324
Kunke, Bob, 328

L

La Cucaracha, 28
La Margarita, it is here!, 121

La Margarita's riches, 188
lab, 325
Lake Michigan, 7
large fine, 328
large silver bowl, 251
large stream anchor, 182
Larry, 128, 147, 157, 159, 194
last big layoff, 324
last visit to the Atocha and Margarita, 329
lead doubloon, 140
lead sheathing, 146
Leah, 164, 171, 173, 181, 220, 291, 319, 332
Leah and Ted, 312
leave of absence, 324
LeClaire, R.D., 94, 132, 172, 238, 277
Lee, 228, 331
legal status, 303
LeMare, Jacques, 303
lightning, 83
lignum vitae cross, 258
limited artifact recoveries, 96
Linda, 331
line-of-sight, 83
Linzee, Claudia, 196
Little, Ed, 172, 208, 220, 236
Littlehales, Nikki, 228
lobster pot buoys, 53
local characters, 49
log entries, 91
London, Jill, 311, 320
lone palm tree, 58
long hours underwater, 154
long lost artifacts, 332
looking for the Atocha, 59
loose coins, 269
LORAN, 113, 143, 144, 166, 178, 302, 328
LORAN-C, 53, 58, 112
LORAN-C's coordinates, 60
Lorrie, 12
lost the trail, 235
lots of artifacts, 124
Lotus, 5
Lower Keys, 17
luck, 148
Lyon, Dr. Eugene, 67, 90, 98, 249

M

machetes, 300
mag, 64, 70
mag boat, 203
mag fish, 63, 64
mag hit, 192

mag hits, 217
Mag is a drag...digging's a blast!, 70
mag surveys, 69
magging, 90, 113, 176
magnetic field, 64
magnetometer, 91, 96, 114, 223, 266, 279,
magnetometer survey, 300
Magruder, 265, 283-332
maiden voyage, 172
mailbox, 110
mailboxes, 26, 43, 225
main artifact area, 124
main pile, 98
Main Pile, 183-184, 246, 249,
254-266, 274, 296, 298

main pile all over again, 151
Main Pile ballast dumps, 296
Main Pile feature, 254
Main Pile site map, 248
Malcom, Corey, 273
manifested gold, 332
mapping benchmarks, 243
mapping procedure, 249
Marathon, 15,
Margarita, 54, 99, 106, 107, 119, 126,
166, 189, 318

Margarita Islands, 107,
***Margarita* Main Pile**, 129, 131-133,
143-146, 157-159, 178, 179, 183,
194, 198, 242, 277, 321

Margarita site, 143
Margarita survivors, 106, 110, 183
Margarita's artifact scatter, 159,
Margarita's primary scatter, 193,
Margarita's wreckage trail, 153, 163
Margaritaville, 248
Mariel, 133
marine surveyors, 166
Marisha, 227
marker buoys, 142
Marquesas, 59
Marquesas Islands, 119, 128
Marquesas Keys, 54, 57, 76, 154, 326
Marquis de Cadereita, 103
married, 222
mass of coins, 124
Massachusetts, 224
massive ribs, 262
mast hole, 139
master chart, 208

Mathewson, Duncan, 98, 130, 131, 171,
220, 238, 244, 249, 253, 269, 275, 291, 302

Matrocci, Andy, 208, 240, 277, 291
Maurice, 55-88
McGill, Ted, 10, 19
McHaley, Bleth, 130, 195
Mel, 217, 265
Mel and Deo, 39, 86
Mel died, 331
Mel Fisher, 70
Mel Fisher Maritime Museum Society, 225, 275,
328

Mel, Deo, Ted, Leah, 246, 249
Melian, 115, 121, 135, 169, 178, 188, 189,
264, 330

Melian, Francisco Nunez, 2, 108
Memorial Day, 236, 260, 263, 275, 296
metal bars, 179
metal box, 63
metal detector, 160, 164
metal detectors, 130, 144
Meyer, Mike, 242, 258
Miguel era, 320
Miguel, Leah, 40, 89, 94, 280
Miguel, Ted, 23
Milky Way, 228
Milwaukee, 6
mining claim, 86
missing pay, 41
Molinar, Demestones, 136
Momo, 136, 137, 144, 149, 161-162, 179, 196,
217, 279

Momo being lucky, 144
money, 46, 209
money chain, 128
money issues, 321
Monroe County Sheriff, 289
Moody, Frank, 132
Moore, Bill, 258
Moore, Michael, 210
Moore, Mike, 268
Moorish design, 161
Moran, Bob, 172, 175-176, 190, 217,
220-221, 227, 244, 330

Moran, General Juan de Lara, 107
Moran, Juan, 103
Mother Lode, 91, 98
mud area, 224

mud zone, 192, 229
Museum, 279
musket balls, 180

N

naos, 103
National Geographic crew, 151
National Geographic documentary, 9, 18, 26, 30, 39, 71, 114, 116

National Geographic film team, 255
National Geographic guys, 148
National Geographic special, 7, 278
National Geographic staff, 130
National Geographic theme song, 65
national marine sanctuary, 325
Navy, 33
Navy Administration building, 207
need-to-know basis, 60
nepotism, 227
nesting weights, 180
new bosses, 283
new detectors, 145,
New Spain Fleet, 102
New Town, 17, 268
New World, 101
New Year, 293, 322
nightly anchorage, 143
Nikon camera, 95
Nine Cannons, 67, 204, 205, 209, 210, 216, 224, 231, 260

no-decompression dive tables, 260
non-ferrous items, 63
North Anchors, 224, 260
North Cannon/Anchor, 183, 184
North Carolina, 303
north tower, 62
Northwind, 71, 98, 137, 254, 278
Northwind sinking, 135, 242
November 2nd, 74
nudist, 170
Nuesta Senora De Rosario, 104
Nuestra Senora de Atocha, 103
nuggets, 162

O

Olaf, 35
old steam engine, 305
Old Town, 17, 161, 198

olive jars, 268
Oops…, 73
Operations Manager, 31, 327
Orlando, 15
Ouija, 235
over the target, 65
overlays, 249
Overseas Highway, 16
ownership of 1715 wrecks, 223

P

Pam, 170
Panama, 102, 104,
pancreatic cancer, 312
panic, 86
Papal seal, 183
paper mache object d'art, 71
Patch Reef, 113, 176, 216, 217, 225
Patch Reef was finally disproved, 236
Patience Pays, 289
Patricia, 113
pay the piper, 284
paychecks, 199, 217
payroll, 208
Pearl divers, 107
Pennington, Craig, 191
Pennsylvania, 31
percentage, 109, 110, 135, 277
Peru, 104
Philippines, 101
Phillips, Greta, 301
photo mapping, 135
photo-mosaic, 256
piece-of-eight, 73, 94, 95, 125, 150, 242, 284
Pier House, 112, 129, 132, 218
pilot of the Guard Fleet, 106
pilot's license, 190
pin ball theory, 217
Pirates, 20
pistol, 48, 120
plastic milk crates, 243
plastic wrapped bales, 198
plateau of soft corals, 97
plotter, 53
Plus Ultra, 172-182, 188, 192, 194, 217, 226, 231
Plus Ultra mag hits, 236
Plus Ultra sidescan sonar printout, 239
pockets, 169
political adventure, 268
pool, 46
Pope Gregory XIII, 184

popping noises, 87
popular research booklet, 324
Portobello, 103, 104
pottery, 274
pottery shards, 119, 233
Preater, Stuart, 186
precision was stunning, 167
prima donnas, 212
prison ship, 201
professional pride, 197
promotion, 93
prop blades, 77
propwash deflectors, 26, 43, 63
psychic powers, 41
public hearings, 327
publicity, 270
Puerto Rican Senate, 268
Puerto Rico, 266
pumped dry, 83
Put away the charts, we've found it!, 241
put on her red dress, 164
PVC grids, 242

Q

quarry men, 116
quartz cannonball, 131
Queen Sophia, 167
Quicksands, 53, 69, 111, 173, 193, 211, 212, 225, 229

R

Raabe, Pam, 220
radio call, 83
Radio Shack, 113
railroad track, 15
raining emeralds, 274
ran out of money, 323
Randolf, Gary, 327
rate of recovery, 179
reciprocal course, 64
recover treasure, 244
religious artifacts, 215
remote sensing, 175
remote sensing equipment obsolete, 325
replica of a Spanish galleon, 18
Ribicoff Pegasus, 137, 256
Ribicoff, Dimitri, 137
Richard, 320
Rick, 214
Rick Lueneburg, 313
Rick's, 112
rifle, holding a , 88
Righthand man, 330
Rio Mar, 223
Robertson, Cliff, 270, 278, 331
Robinson Caruso, 55
roller coaster, 324
Rosario, 106, 107
run down the *Virgalona*, 88

S

Saba Rock, 229-238, 274
safety cages, 156
safety line, 58
salvage auditor, 169
Salvors Inc., 279
Salvors' Atocha finds, 189
Samba Rock, 224
Sanctuary boundaries, 326
Sanctuary helicopter , 326
Sanctuary lawsuit, 326
Sanctuary officials, 332
Sandy Point, 223, 318
Santa Fe de Bogotá coin, 205
Santa Margarita, 20, 103, 109, 127, 133, 135, 203

sardine tins, 214
Save the Bales, 198
scattered ballast, 124
scattered over twelve miles, 260
Schooner Warf Bar, 112
scrapped, 321
scruffy houseboat, 89
scuba tank refilling system, 83
scuttlebutt, 282
Sea Hunt, 6
sea sled, 182
seaborne chess game, 88
Seaborne Ventures, 92, 93
search coil, 66
Sebastian, 202
secondary scatter, 159
second-class outsiders, 212
Seeker, 86
selected by dartboard, 91
semi-automatic weapons, 294
September 6th, 106
serious business, 79
set–up process, 143
Seven Mile Bridge, 17

seven miles, 260
seventy-seven gold bars, 258
sharks, 310
ship's construction details, 90
shipwreck had been found, 245
Shoreline 21, 299, 301
shrimp dock, 52
shrimp processing plant, 37
side-scan sonar, 122, 182, 317
silver bar, 121, 150, 164, 184, 210
silver bar area, 249
silver bars, 124, 127, 133, 183, 240-247, 280, 286-289

silver bowl, 248
silver bowls, 269
silver coins, 170, 250, 279, 288
silver coins everywhere, 125
silver pitchers, 251, 273
silver plate, 269
silverware, 269, 274
Simonton Street, 136
simple excavation, 145
Sinclair, Jim, 249, 251
Singleton Seafood, 34, 52
sinking, 208
slaver, 106
sled, 165
Sloppy Joe's Bar, 30
sluice box, 116
small gold coin, 139
small loop detector, 214
Smith, Debbie, 184, 186
Smith, Jedwin, 236
smuggled bars, 211
Sneaker, 86
solution hole, 128, 145, 146, 321
Southwind, 137, 228
Spanish accounts, 96
Spanish galleon, 76
Spanish lake, 101
Spanish shipwreck sites, 6
spar buoys, 144, 260
special plans, 265
spiderweb, 74
spikes, 224, 269
split shot, 182
Spring of Whitby, 223
St. Johns, 299
State of Florida, 158, 217, 218
State of Florida archives, 112
Steel Island, 238
steel wreckage, 62

stepped, 139
Steve, 62
stock in the company, 153
stomach cramp, 296
stopping payments, 293
stormy company matters, 291
Stuart, 128, 150-157, 162-165, 170, 185-191, 194, 202-210, 217

submerged cultural resources, 325
summer tropical wave, 45
Summerland Key, 133
Super Bowl-type rings, 278
survey data report, 112
survey of the Quicksands, 69
survey project, 327
surveys, 96
Susan, 230
swagger, 85
Swanson, Christian, 268, 275
Sweat of the sun, tears of the moon, 152
Switt, Lorretta, 278
Swivel Gun, 237-239, 260-264, 275
Swordfish, 23, 26, 36-52, 110, 177, 188, 197, 241, 265, 292

Swordfish's stern, 58
swords, 150
symbiotic relationship, 31
system of buoys, 178

T

Taffi, 197, 228, 279, 325, 332
tagged, 72
tangle of gold money chains, 163
tardy arrival, 253
tax laws, 279
tax writeoffs, 282
Tears of the Moon, 121
Ted and Leah, 227, 236, 268, 279, 314-315
Ted division disagreement, 277
Ted Miguel died, 316
Ted stomach problems, 312
Ted terminated, 291
Ted, dislike of, 111
Ted's intimidating nature, 42
Ted's nickname - Capitana, 149
Tennessee Williams Theater, 278
terminated, 294
Tern, 164, 175-179, 188, 207
Terri, 303

test pits, 249
The World's Greatest Treasure Hunter, 197, 267, 306, 330

The Beef, 196
the Black Sheep Squadron, 197
The Cannons, 98
the gold line, 213, 215
The King of the Conch Republic, 330
The **Main Pile**, 240
the old ship place, 17
The Silent World, 6
theodolite, 62
theory, 217
Thirteen, 95
thousands of anomalies, 228
thousands of coins, 263
Thousands of nuggets, 162
Thrasher, Chuck, 180, 186
three weeks notice, 221
tidal current, 58
Tierra Firma Fleet, 102
Time Magazine, 248
time to invest, 239, 243
Tom, 243, 244, 289
Tony, 304-329
tough spot, 244
tow a diver, 143
towing job, 209
toys, 294
transponder, 166
treasure hunting, 91
Treasure Salvors' conservation lab, 73
Treasure Salvors' diver roster, 51
Treasure Salvors' location, 18
Treasure Salvors' only detector, 74
Treasure Salvors, 10, 41, 279
Treasure Salvors salary, 12
Treasure Salvors, employment at, 21
tribbits, 254
Trisponder, 166
Trotta, Vince, 284, 292
Truman Annex, 133, 207
tugboat, 209
twenty-eight pounds of gold, 163
two escudo gold coin, 319
Two Friends Bar, 112
two hundred silver bars, 263
Two large galleon anchors, 224

U

U.S. Federal Marshall, 133

ultimate possession, 89
ultimatum, 31
uncertain paychecks, 322
unchallenged title, 218
unconventional family, 197
uncut emerald, 280, 287, 262
Unit 10, 316, 332
Unit 13, 316
Unit One, 316
unlighted towers, 63
used nuclear test site, 30

V

Vargas, 2, 107, 110, 159, 264, 330, 108
Vargas, Gaspar de, 104, 106
Vero Beach, 197
VHF radio, 53, 79, 82
Vieques, 267, 268
Vince, 293
Virgalona, 34, 41-42, 111, 118, 123, 144, 172, 180, 217, 238, 242, 279

Virgalona crew, 136

W

W-2, 286
W-2 tax form, 284
Wagner, Kip, 222, 267, 310
walk out, 31
wall of barracudas, 62
Wally, 28, 53, 63, 91, 140, 147, 158, 161
Warbird Museum, 331
Wareham, Greg, 240, 293
Waterfront Market, 292
weapons, 47
Weller, Bob and Margaret, 224
Whaler, 309
Whaler's destiny, 308
White, Curtis, 238
Whydah, 224
wildcat projects, 322
wildcatting, 91, 180, 267, 298, 303, 320
Wildlife Rescue of the Florida Keys, 326
wildlife rescue organization, 324
winch, 78
winch incident, 79
Wisconsin, 95, 175
Women for Sail, 311, 320-325
wood, 124
wood structure, 249

wooden box, 269
World War II Avenger, 289
World War II wrecks, 223
WWII fighter trainer, 331

Z

zigzag course, 60
Zitver, Geoff, 322, 323
Zwierson, Maryann, 315

The Author

Syd Jones joined Mel Fisher's **Treasure Salvors** team in 1979 and was involved in the major discoveries on the *Santa Margarita* and *Nuestra Senora de Atocha* sites, logging more than 7,000 hours underwater. During his 18 years, he captained several of the survey and recovery vessels, ultimately becoming the company's operations manager. Syd was also involved in the *1715 Fleet* salvage operations off the east coast of central Florida as well as numerous shipwreck projects in the Caribbean.

Syd and his wife KT left Treasure Salvors in 1997 to restore and fly a WWII bomber and fighter aircraft in central Florida. They were later employed to help start the Pacific Aviation Museum in Pearl Harbor, Hawaii. They continue to look for new adventures.